Handbook of Charcoal Making

T0234941

Solar Energy R&D in the European Community

Series E:

Energy from Biomass

Volume 7

Solar Energy R&D
in the European Community

Series E   Volume 7
Energy from Biomass

Handbook of
Charcoal Making

The Traditional and Industrial Methods

by

WALTER EMRICH

D. Reidel Publishing Company

A MEMBER OF THE KLUWER ACADEMIC PUBLISHERS GROUP

Dordrecht / Boston / Lancaster

for the Commission of the European Communities

Library of Congress Cataloging in Publication Data

Emrich, Walter.
     Handbook of charcoal making.

          (Solar Energy R&D in the European Community. Series E, Energy from
     biomass; v. 7)   (EUR; 9590)
          Includes bibliographies.
          1.   Charcoal.   I.   Commission of the European Communities.
     II.   Title.   III.   Series: Solar energy R & D in the European Community.
     Series E, Energy from biomass; v. 7.   IV.   Series: EUR; 9590.
     TP331.E45   1985   662'.74   84-27630

     ISBN 978-90-481-8411-8

Publication arrangements by
Commission of the European Communities
Directorate-General Information Market and Innovation, Luxembourg

EUR 9590
© 1985 ECSC, EEC, EAEC, Brussels and Luxembourg
Softcover reprint of the hardcover 1st edition 1985

LEGAL NOTICE
Neither the Commission of the European Communities nor any person acting on behalf of the
Commission is responsible for the use which might be made of the following information.

Published by D. Reidel Publishing Company
P.O. Box 17, 3300 AA Dordrecht, Holland

Sold and distributed in the U.S.A. and Canada
by Kluwer Academic Publishers,
190 Old Derby Street, Hingham, MA 02043, U.S.A.

In all other countries, sold and distributed
by Kluwer Academic Publishers Group,
P.O. Box 322, 3300 AH Dordrecht, Holland

PREFACE

We are happy to introduce the Handbook of Charcoal-Making, a comprehensive survey written by a competent expert with international experience. The book was prepared by the Commission of the European Communities in the frame of its R + D programme on biomass.

In the European Community today the biomass option is only little developed: a huge resource is waiting for use.

Actually, there is ample scope for biomass utilisation as it bears promise in some of the vital sectors of modern society. Development of indigenous and renewable energy sources, creation of new employment, recycling of wastes and improvement of the environment, restructuring of European agriculture, development of the Third World, they are all concerned.

It is important to note that the exploitation of the biomass resource is largely related to its conversion into a marketable product. However, as many of the conversion technologies are not yet well established or need improvement, R + D is more than ever the critical pathway to get access to the benefits of biomass utilisation.

In the European Communities' R + D programme, thermal conversion of biomass is developed with priority. Gasification as well as pyrolysis development projects are being supported by the Commission in European industry and universities.

Pyrolysis is particularly attractive because the conversion products charcoal and pyrolytic oil are very convenient in use, technologies are relatively simple and projected pay-back times favourable.

Charcoal making is just the simplest and oldest form of pyrolysis. Charcoal is already a market product and plays an important role in the energy consumption structures of most developing countries.

As modern literature on charcoal is scarce, this book will first of all serve the purpose of a review book of the state-of-the-art. Furthermore, it is essential as a reference book for future R + D in view of technical improvements and new processes of charcoal making and pyrolysis in general.

I take this opportunity to thank Dr. Walter Emrich for having accepted the Commission's invitation to write this book. I also thank Mr. L. Crossby and Mr. J.F. Molle for reviewing the manuscript.

I wish the book great success.

Dr. W. Palz
R + D Programme Biomass
Commission of the European Communities

CONTENTS

PREFACE   v

LIST OF ILLUSTRATIONS   xi

FOREWORD   xv

Chapter 1
HISTORY AND FUNDAMENTALS OF THE CHARCOAL PROCESS
1.1   Charcoal-Making from the Beginning until   1
           the Present Day   1
1.2   Theory of the Carbonization Process   5
1.3   Heating Systems for Charcoal Plants   8
1.4   Properties of Carbonization Products   11
1.4.1   Charcoal   12
1.4.2   Pyrolysis Oil   16
1.4.3   Process Gas   17
           References   18

Chapter 2
TRADITIONAL METHODS OF THE SMALLHOLDER CHARCOAL-
MAKER   19
2.1   Charcoal Pits and Earthmound Kilns   20
2.1.1   The Charcoal Pit   24
2.1.2   The Earthmound Kiln   27
2.1.3   The Earthmound Kiln with Chimney   33
2.1.4   The Earthmound Kiln with Tar Recovery   35
2.2   Charcoal-Making with Portable and
           Movable Kilns   38

2.2.1	The Tongan Oil Drum Kiln	39
2.2.2	The Philippines Kiln	42
2.2.3	The Black Rock Forest Kiln	45
2.2.4	Sectional Metal Kilns	48
2.2.5	The Carborion Kiln	54
2.3	Concrete and Brick Kilns	56
2.3.1	The Missouri Kiln	60
2.3.2	Cinder Block Kilns	66
2.3.3	The Schwartz and Ottelinska Furnaces	74
2.3.4	The Brazilian Beehive Brick Kiln	77
2.3.5	The Argentine Kilns	88
2.4	Kiln Designs for Waste Conversion	99
2.4.1	The Carbo-Gas Retort	101
	References	104

Chapter 3
CONCEPTS AND TECHNOLOGY FOR THE INDUSTRIAL
CHARCOAL-MAKER

		107
3.1	Equipment for Charcoal Plants with By-Product Recovery	107
3.1.1	The Forerunners of Modern Charcoal Equipment	109
3.1.2	Retort Technology	116
3.1.2.1	The Wagon Retort	117
3.1.2.2	The Reichert Retort Process	120
3.1.2.3	The French SIFIC Process	123
3.1.3	Charcoal Technology for The Carbonization of Biomass	129
3.1.3.1	Generalized Flow Diagram	129
3.1.3.2	The Multiple Hearth Furnace	133
3.1.3.3	The Fluid Bed Carbonizer	136
3.1.3.4	The Vertical Flow Converter	139
3.1.3.5	The Enerco Mobile Pyrolyser (Model 24)	143
	References	147

Chapter 4

TECHNIQUES FOR RECOVERING COMMERCIAL PRODUCTS FROM
PYROLYSIS OIL   148
4.1   Pyrolysis Oil Recovery   149
4.2   Crude Acetic Acid and Acetone Recovery   150
4.3   Recovery of Methanol (Wood Spirit)   152
4.4   Processing of Charcoal Tar   154
4.5   Concluding Remarks   159
           References   161

Chapter 5

RAW MATERIALS SUPPLY   162
5.1   Supply from Fuelwood Plantations   165
5.2   Agricultural Resources   166
5.3   Transport and Preparation of Raw Materials   168
5.3.1   Key Factors in Wood Supply   169
           References   176

Chapter 6

END-USE MARKETS FOR CHARCOAL AND CHARCOAL BY-PRODUCTS   178
6.1   Charcoal as Household Fuel   178
6.1.1   Lump Charcoal   178
6.1.2   Charcoal Briquettes   179
6.2   Charcoal as Fuel for Industry   180
6.3   Charcoal in Metal Extraction   182
6.4   Activated Charcoal   183
6.4.1   Synopsis of Industrial Active Carbon Markets   185
6.5   Speciality Markets for Charcoal   187
6.6   Charcoal for Producer Gas   187
6.7   By-Product Utilization   189
6.8   Synopsis of Major Uses of Charcoal and
           By-Products   193
6.9   Charcoal Costs and Fuel Prices   195
6.10   Packing and Shipment for Export/Market
           Strategy   200
6.11   World Production   203
           References   207

Chapter 7
PLANNING A CHARCOAL VENTURE AND SELECTION OF EQUIPMENT 208
7.1   Planning of Projects   211
7.2   Selection of Charcoal Equipment   214
7.3   Conclusions   220
        References   222

Chapter 8
CHARCOAL BRIQUETTES AND ACTIVATED CHARCOAL
MANUFACTURING   223
8.1   The Briquetting Process   223
8.1.1   Simple Briquetting Equipment   227
8.2   The Activated Charcoal Process   228
        References   233

Chapter 9
SAFETY PRECAUTIONS AND ENVIRONMENTAL CONSIDERATIONS   234
9.1   Safety in Charcoal Operations   234
9.2   Safety Devices and Equipment   236
9.3   General Safeguarding of Charcoal Plants   237
9.4   Precautions for Charcoal Storage   238
9.5   Environmental Considerations for the
        Charcoal-Maker   239

Chapter 10
CHARCOAL LABORATORY WORK   243
10.1   Analysis   244
10.2   Bench-Scale Carbonization Tests   251
        References   253

APPENDICES   254
Appendix 1   Case Studies   255
Appendix 2   Energy Distribution Diagram   265
Appendix 3   Addresses of Consultants, Institutes,
              and Equipment Suppliers   268
Appendix 4   Conversion Tables   275

LIST OF ILLUSTRATIONS

Figures

1, 1a	Temperature diagrams of dry distillation	6
2	Temperature distribution diagram (continuous process)	7
3	Heating systems	9
4	Classification of charcoal processes	10
5, 5a	Carbon content and higher heating value. Moisture content and net heating value	23
6	A charcoal pit	26
7	A small earthmound kiln	28
8	A large earthmound kiln with centre firing	31
9	An earthmound kiln with chimney	34
10	An earthmound kiln with pyrolysis oil recovery	36
11	The Tongan oil drum kiln	40
12	The Philippines oil drum kiln	43
13	The Black Rock Forest kiln	46
14	Handling the kiln with a derrick	46
15	A battery of four Black Rock Forest kilns in operation	47
16	A portable metal kiln	50
17	Air inlet channels at the bottom of a portable metal kiln with wood grate	50
18	The Carborion kiln	55
19	The Missouri charcoal kiln	58
20	Plan and elevation of the Missouri kiln	59
21	A cinder-block charcoal kiln	67
22	Typical masonry units for block-type charcoal kilns	70
23	Detail of the thermocouple assembly on the lengthwise centreline of a cinder-block kiln	72
24	The Schwartz charcoal furnace	75
25	The Ottelinska furnace	75

26	Improving the Schwartz system by installing "calorifères"	76
27	The Brazilian beehive brick kiln	79
28	The slope-type beehive brick kiln	80
29	Beehive fire brick kiln with external heating	81
30	A charcoal production centre	86
31	Half orange kiln with straight jacket	91
32	Carbo-Gas retort (twin unit)	100
33	Carbo-Gas retort plant with charcoal gas recovery for commercial use	100
34	The Carbo furnace	111
35	The Bosnic charcoal plant	113
36	A smaller Bosnic plant with interchangeable retorts	114
37	The wagon retort plant	118
38	The Reichert retort process	122
39	The French SIFIC retort process	124
40	Side view of the CISR Lambiotte plant	127
41	Generalized flow diagram of the rapid pyrolysis process	131
42	Cross-section of a multiple hearth furnace	134
43	The fluid bed carbonizer. Generalized diagram	137
44	The vertical flow converter	140
45	The ENERCO model 24 pyrolyser	145
46	A charcoal plant with pyrolysis oil refinery	153
47	Recovery of commercial products from pyrolytic tar	155
48	A wood dryer for continuous operation	174
49	The integrated carbonization concept with four carbonizers	212
50	Simple charcoal briquetting press	226
51	Activated carbon plant for manufacturing of pellets or granular active carbon	231
52	Apparatus for bench-scale dry distillation	252
53	The energy distribution diagram	266

Photos

1   Small earthmound kiln in Ghana one hour
        after lighting.   30
2   Discharging charcoal from same kiln
        two days later.   30
3   A Missouri kiln. The shell is dangerously
        cracked as a result of faulty operation.   63
4   Side view of kiln with two smoke pipes and
        air inlet holes at the bottom   63
5   Charging the beehive brick kiln.   85
6   A Brazilian beehive brick kiln in full
        operation.   85
7   The Argentine half-orange kiln. The operator
        is closing the gate after charging the kiln.   90
8   Small half orange kiln (7 m^3)   92
9   Charcoal trainees in Kenya constructing
        a half orange kiln with straight jacket.   92
10   Charcoal trainees igniting the kiln with
        a shovelful of glowing charcoal.   93
11   Charcoal trainee brushes over leaks.   93
12   A Lambiotte reactor.   128
13   Model of a vertical flow converter
        charcoal plant.   141
14   A charcoal briquetting press.   225
15   Pillow-shaped charcoal briquettes.   225
16   A rotary kiln for activation of charcoal
        in the Philippines.   232

FOREWORD

Owing to the widespread use of cheap fossil fuels and natural gas in industry, household charcoal has been somewhat neglected during recent decades. The development of new and improved charcoal techniques has nevertheless been advancing during this period, unknown to outsiders.

Comprehensive charcoal literature has not appeared since the late nineteen-forties; in particular, there have been no publications concerned with industrial charcoal-making. Some of the literature cited in this book exists only in specialized collections. Occasionally the public has learned about the achievements of companies active in charcoal production or equipment manufacturing, particularly in the carbonization of biomass and the formulation of long-burning charcoal fuels, but overall there has been an inadequate flow of information to potential users.

The Commission of the European Communities, Directorate-General for Science, Research and Development intends to close the information gap by publishing this handbook. However, a handbook cannot be expected to reach all the innumerable small-scale charcoal-makers, distributors and users, especially in developing countries, who do not normally acquire knowledge of improved techniques from books. At this level, information should be disseminated directly by government agencies or where appropriate, through internationally sponsored development projects.

The author has been engaged as consultant and design engineer in the charcoal and active carbon industry for more than twenty years. He has also worked on assignments as research and plant manager of charcoal and active carbon plants. During these years he became aware, through numerous contacts with Governments, Ministries of Planning and private entities that two factors frequently prevent or obstruct the promotion and realization of efficient projects:

- inadequate knowledge of the state of the art

- lack of the experience needed to develop charcoal projects.

The author's major concern in this handbook is to draw the attention of all persons involved in energy project planning to the fact that new and improved charcoal techniques are able to convert forestal and agricultural wastes and residues into energy. In countries which abound with these reserves the modern charcoal-maker can make an important contribution to the household fuel programme of his country.

For more than a thousand years, charcoal has been made from whole trees; it is time for everyone to accept recent advances in a very old industry and to adopt new ways. We should always bear in mind:

THERE IS NO WASTE IN THE WORLD
WASTE IS AN ENERGY RESERVE

Therefore, let's use it.

The author would like to express his gratitude to the numerous organizations which have contributed valuable data. Among these are: the United Nations Industrial Development Organization (UNIDO), the Food and Agriculture Organization in Rome, and the Barbeque Industry Association in the U.S.A.

Last but not least, the author would like to thank the many charcoal producers and equipment suppliers who have volunteered updated proprietary information.

WALTER EMRICH

Neu-Isenburg, September 1984

Chapter 1
HISTORY AND FUNDAMENTALS OF THE CHARCOAL PROCESS

1.1 Charcoal-Making from the Beginning until the Present Day

Prehistoric finds, dating back six thousand years, have
shown that arrow-heads were attached to their shafts by
employing wood tar, a material then obtainable only by the
charring of wood. Although we may never know when man first
made charcoal, extensive investigation has proved that in
Europe charcoal-making had already become an important
industry for the recovery of iron and other metals from
their ores around 1100 BC.

The Roman historian Plinius describes in his famous
Historia Naturalis (1) the method of embalming and preparing
bodies for burial in Egypt, in which the watery condensate
of the charring process was used as the preserving agent.
The more viscous parts of the condensates like tar and pitch
found applications as house paints, in the then flourishing
shipbuilding industry, and for caulking and sealing wooden
barrels.

We may assume that charcoal was made mainly in open pits
at that time, but with very low yields. Later some
improvements were made by introducing the earthmound kiln,
which is still common in many developing countries today.
Although this technique is not at all satisfactory in so far
as the energy balance of the charcoal process is concerned,
it can be considered as an industrial step, because by
simple alterations one is able to collect some liquid
by-products.

In 1635, the renowned chemist Rudolf Glauber (2)
discovered acetic acid as an essential component in the
condensate of the charcoal process. Immediately his
discovery drew the attention of the chemical industry to the
charcoal process as an available source of raw materials.

However, the earthmound kiln method did not permit the production of the desired quantities of condensates (now termed "pyrolysis oil"). Almost a hundred years were to pass before the Swedish engineer Nordenschoeld (3) and the German technician Reichenbach (4) designed commercial retorts by which the energy balance of the process could be improved considerably and the by-product yield was raised sufficiently. The new "charcoal furnaces" spread throughout most European countries and from there they were introduced to the U.S.A. and Canada. The Swedish-German technology was very successful and was only replaced by large-scale equipment and improved techniques in 1850.

The growing demand for steel by industry in general, and the enlarging capacities of the chemical industry, brought about an unprecedented upsurge of charcoal and the liquid by-products. Small and large-scale charcoal plants mushroomed not only in Europe but also in North America. Not only the number of production sites multiplied, but also the capacity of the plants. This again emphasized the need for better and more efficient technology.

Technical development was focused on continuous processes and techniques to utilize the total accessible energy contained in the raw material. Sophisticated combustion systems were invented to generate heat or electric power from the non-condensable gases of the process.

Despite these achievements (which resulted in better charcoal quality), the steel industry and metallurgy gradually turned to new resources for their blast furnaces: refined bituminous coal, coke and lignite began to compete with charcoal. The chemical industry also found that other sources were more rewarding, namely bituminous coal and later fossil oil and natural gas.

The already foreshadowed decline of the once-blossoming charcoal industry became visible after the First World War. Although a great number of plants had to be shut down, the more advanced and well managed plants survived the crisis.

Then, with the economic recovery of the industrialized countries, the demand for charcoal products revived. Countries lacking in natural resources and therefore heavily dependent on imports for all kinds of goods, curtailed their spending of foreign currency and tried to exploit their own reserves. Naturally, under these conditions, their forests played an important role. Consequently the charcoal-makers experienced a new peak within their exciting lifespan.

When the Second World War began, the charcoal industry had in many countries a somewhat tactical task to sustain the war machinery. The products provided by carbonization plants were manifold and far exceeded the expectations of industrialists and businessmen a century ago.

Charcoal came to be utilized not only in metallurgy but also in chemical processes and the fabrication of plastics, it was the essential raw material for activated carbon with its numerous outlets in the filter industry, pharmaceuticals, catalysts, pollution control, etc. Pyrolysis oil provided a basis for the production of organic acids, methanol, aldehydes, acetone, creosostes, tars, etc.

In 1945, at the end of the war, the economy of many European countries had come to a complete standstill. Again the charcoal industry was hard hit. Survival meant concentration and enlarged capacities, in other words automatization.

During the post-war years, a new technology was commercially proved, the so-called "rapid pyrolysis", which was first operated in the U.S.A. It had two major advantages over previous processes: first, the carbonization cycle (residence time of the feed) was shortened dramatically, thus making more profitable use of the invested capital; secondly, the conversion of small raw material particles was made possible.

Rapid pyrolysis is one of the most important achievements of charcoal technology. Whereas, until then, the only feed for the charcoal plant consisted of pile wood or wood logs cut to size, the new techniques permitted the

-3-

utilization of raw material up to a few millimetres in diameter. This opened the door to a new category of raw material resources until then untapped and neglected, namely agricultural and industrial wastes and forestal residues. One is tempted to say that these reserves, which abound in developing countries, are almost unlimited. Their use to provide household fuel, industrial charcoal and chemical raw materials could make an important contribution to the conservation of other natural resources, especially the highly endangered forests.

When early man invented charcoal, its only use was as a reductant to obtain metals from their ores. Numerous applications have been added since then, and new markets have opened their doors to the effective charcoal-maker. Soon after the Second World War, people in industrialized countries rediscovered the excellent properties of charcoal for barbequeing. This specialized sector of the charcoal industry has now become a major economic factor, comprising also thousands of manufacturers of grills and utensils.

Sufficiently accurate data on the production and consumption of charcoal and the by-products are difficult to obtain. Besides charcoal, fuelwood is the main energy carrier with which people cook their daily food in developing countries. The world consumption of fuelwood, per capita, including charcoal was estimated at 0.37 m^3 in 1978 (5). However, in the developed world the per capita usage reached only 0.13 m^3, compared with 0.46 m^3 in the developing world.

The United Nations Conference in Nairobi, 1981, (6) concluded that approximately 2,000 million people are utilizing firewood and charcoal for cooking and heating. For several hundred million of them, it is already impossible to find a sufficient supply of firewood because of the ruinous exploitation of forests. Here the modern charcoal technology with high-grade energy recovery systems can find a new task. By tapping the vast waste reserves of the world, the charcoal industry can make one of its most important contributions to mankind by helping to provide for the energy needs of the future, especially in all developing countries.

1.2 Theory of the Carbonization Process

Carbonization or dry distillation takes place when any organic matter is raised to a high temperature (i.e. above 180° C) under strict exclusion of oxygen or under controlled minimal air intake. Essentially the process of carbonization follows a general temperature scheme:

- between 100° and 170° C all loosely bound water is evaporated from the raw material.

- between 170° and 270° C gases develop (off-gas), containing carbon monoxide (CO), carbon dioxide (CO_2), and condensable vapours, which form pyrolysis oil after scrubbing and chilling.

- between 270° and 280° C an exothermic reaction starts, which can be detected by the spontaneous generation of heat and the rising temperature. At the same time, the development of CO and CO_2 ceases but the quantity of condensable vapours rises.

Once the carbonization process has entered the exothermic phase, no more outside heating is required. The temperature in the retort will climb slowly until it comes to a standstill between 400° and 450° C.

Naturally, this scheme can be applied only if the carbonization or dry distillation is conducted batch-wise. To achieve a higher terminal temperature, the process must be supported with extra heat from outside.

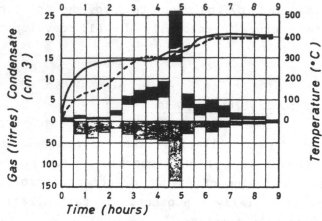

Figure 1 Diagram of dry distillation (7)
For Softwood

Graphs-solid curve: temperature outside retort
dotted curve: temperature at retort centre

Bar diagrams-solid area: combustible gas
white area: CO_2
shaded area: condensate

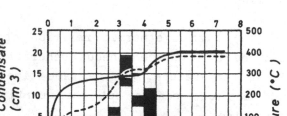

Figure 1a Temperature Diagram of dry distillation (7)
For Hardwood

(After M. Klar, Technologie d. Holzverkohlung)

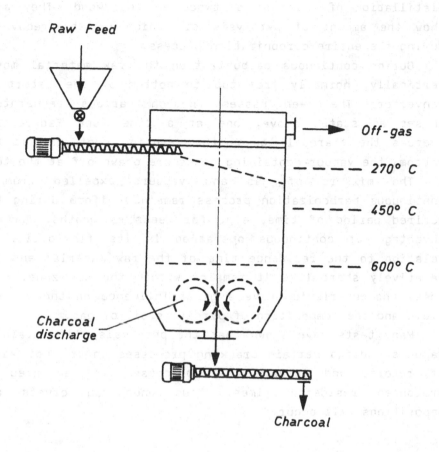

Raw Feed

Off-gas

——— 270° C

——— 450° C

——— 600° C

Charcoal
discharge

Charcoal

Figure 2 _Temperature distribution_
diagram (continuous process). (8)

(From: W. Emrich, Recycling in
Developing Countries)

Figures 1 and 1a are representative of the dry distillation of European softwood and hardwood. They also show the amount of pyrolysis oil which can be recovered during the entire carbonization process.

During continuous carbonization the raw material moves vertically, normally from top to bottom of the retort or converter. The feed passes through various temperature phases as stated above, one at a time (see Figure 2). Whereas the charcoal leaves the reaction container at the bottom, the vapour-containing gases are drawn off at the top.

The mixture of gas and vapours expelled from a continuous carbonization process remains uniform during the desired period of time, e.g. for weeks or months. Another advantage of continuous operation is its flexibility in relation to the residence time of the raw material and the relatively short time it remains within the hot zone. Both these characteristics have a great influence on the process yield and the composition of pyrolysis oil or gas.

Many tests have shown that the pyrolysis oil containing vapours undergo certain cracking processes on the hot walls of retorts and tubes. These processes are enhanced by prolonged residence times, thus undesired crusts and depositions will occur.

1.3 Heating Systems for Charcoal Plants

To start up the carbonization and to maintain higher temperatures, external heating is required. During the long history of industrial charcoal-making, many heating systems have been tried. All were intended to reduce expenses and fuel costs. Only the three basic types shown in Figure 3 have survived and are in general use.

A

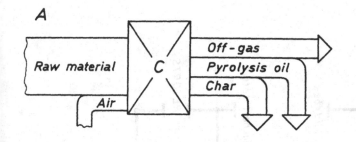

Internal heating by controlled combustion of raw materials.

B

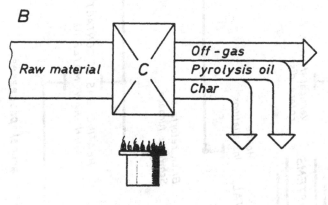

External heating by combustion of fire-wood, fuel oil or natural gas

C

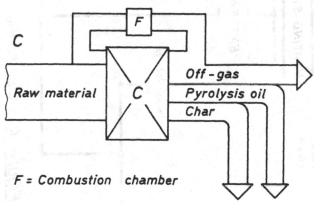

Heating with recirc-ulated gas (retort or converter gas). Hot gases pass through raw material charge.

F = Combustion chamber

Figure 3   Heating systems (9)

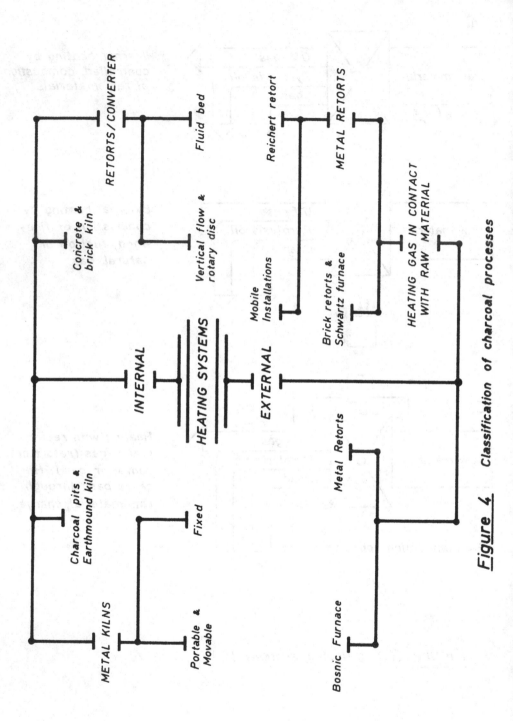

Figure 4 Classification of charcoal processes

Tpye A: Most common system. Part of the raw material is
          burnt under controlled air inlet. The combustion
          heat provides the energy for maintenance of the
          process. Recommendable only in locations where raw
          material prices are low (waste material).
Type B: Retort or converter heated from the outside under
          strict exclusion of oxygen. Fuel can be provided
          from the off-gases.
Type C: Very expensive heating system. Raw material comes
          in direct contact with the hot gases. Charcoal and
          by-product yields are high. Recommendable for very
          large plant capacities only.

     All the systems described are technically and
commercially proved. There will be advantages and
disadvantages which must be considered in making decisions.
Not only the costs for the investment will play a role, but
also the concept which the charcoal planner has in mind.
     For the classification of charcoal equipment, the scheme
shown in Figure 4 is suggested for practical use. Presently,
more than one hundred concepts and methods to make charcoal
are known, all of which can be categorized and tabulated
within this frame.

1.4 Properties of Carbonization Products

     Whether the charcoal or a charcoal derivative can be
regarded as a quality product will depend first on the
results of analytical tests. Chemical and physical
properties are greatly influenced by three factors - raw
material type, process characteristics, and after-treatment.
The latter refers to agglomeration and briquetting
techniques of the solid char, or to distillation and
fractionation methods applied to the pyrolysis oil and
residual process gas.

In any event, it is absolutely necessary during the planning stage of a new charcoal venture to take a close look at the applications and end-use markets of the expected products (see Chapter 6). Specifications may be set by individual consumers and can be obtained from them.

The terms and properties described below are intended to acquaint the reader with the charcoal terminology. They should be read carefully before proceeding to the following chapters.

1.4.1 Charcoal

A precise and authoritative definition of charcoal is becoming increasingly inportant for label regulation and other imminent legislation. Two typical published definitions are:

1. Encyclopedia Britannica (10): "Charcoal is the residue obtained when carbonaceous materials, of either animal or vegetable origin, are partially burned or heated so that tarry and volatile matter is removed; in most cases the residues may be roughly described as impure carbon." (Coal, coke and petroleum coke certainly are not charcoal; but according to this definition they qualify).

2. Encyclopedia Americana (11): "Charcoal, a black, solid, non-lustrous residue, or amorphous carbon, from vegetable or animal substances; or a coal made by charring wood in a kiln or retort from which air is excluded." (Dull bituminous coal is not charcoal but according to this definition it qualifies.)

To resolve this dilemma we propose:

CHARCOAL IS THE RESIDUE OF SOLID NON-AGGLOMERATING ORGANIC MATTER, OF VEGETABLE OR ANIMAL ORIGIN, THAT RESULTS FROM CARBONIZATION BY HEAT IN THE ABSENCE OF AIR AT A TEMPERATURE ABOVE 300 DEGREES CELSIUS.

This definition distinguishes charcoal from coke, which is formed by carbonization of fluid organic matter such as plastic coal or petroleum (when heated, coking coal becomes plastic before it carbonizes). It also distinguishes bituminous coals and lignite from charcoal, because they have not been subject to carbonizing temperatures during their metamorphoses.

The charcoal-maker differentiates between lump charcoal, charcoal fines, charcoal dust, charcoal briquettes, pellets, extrudates, and activated charcoal.

Yield: Expressed as weight of charcoal per unit weight of dry raw material, in percentage. Note that the yield is always applied to the dry material weight.

Specific Weight: Refers to the density of the charcoal, which varies according to the density of the raw material. The density of charcoal can be influenced within a narrow range by the course of the process temperature, in particular by the terminal temperature.

Hardness: A very important coefficient for industrial charcoal. Standard scales have been imposed in some countries; these are normally identical with the hardness degrees of bituminous coal.

Moisture: After the charcoal has left the converter, it vigorously absorbs water from the air up to 6 % of its dry weight. In some continuous operations, the hot charcoal is cooled by a controlled water spray. If treated in this way or stored in the open, the moisture content may be much higher than 5 to 6 %.

Content of Volatiles: If charcoal is heated to 900° under confined conditions, it will lose weight because hydrocarbons and nitrogen are driven out. This weight loss is extremely important to industrial charcoal consumers when defining the utilization properties. In general, the weight loss should not exceed 30 %.

Half-burnt charcoal, or "brands": Product with more than 30% volatiles.

Red Coal: Same as half-burnt charcoal or "brands".

Fixed carbon content: The dry charcoal weight minus the content of volatiles and incombustibles (ashes) is equivalent to the content of fixed carbon, which also determines its fuel value (C_{fix}).

Dead-burnt charcoal: This has a content of fixed carbon which results in difficult ignition of the charcoal.

Ash content: The ash is composed of the natural minerals contained in almost any organic matter and contaminations. The quantity is related to the composition of the raw material mix, e.g. wood branches with a high proportion of bark will give high ash containing charcoal. Charcoal ashes are distinguished by their solubility in water and by chemical analysis.

Sulphur and phosphorus content: The low sum of these substances normally found in charcoals make them especially attractive for use in blast iron furnaces and for metallurgical purposes. The desired value for sulphur is usually below 0.05 % and for phosphorus under 0.03 %.

Heating or calorific value: This depends on the fixed carbon content and will be lowered only by high ash content. In general, heating values range between 6,500 and 7,200 kcal/kg (30,100 KJ/kg), comparable to bituminous coal.

Active surface: The surface of well burnt charcoal is porous. The porosity makes charcoal easy to ignite, reactive in chemical processes, and able to absorb substances and remove them from liquids or gases. The porosity can be measured, and is expressed in m^2/g.

Active or activated charcoal: The porosity or surface area can be enlarged by special activation processes. Industries use gas, steam or chemical activation. The largest surface area which can be achieved in commercial plants measures approximately 1,500 m^2/g, which is close to the area of a soccer field.

Agglomerated and briquetted charcoal: Some industrial applications and major barbeque markets demand char pieces with a particular shape. Known shapes and forms are spheres, cylinders, hexagonals, diamonds, bricks, oblong and pillow-shaped conglomerates and pellets. The constituent parts are: charcoal, binder, additives.

Energy extender, cooking-time extender: These are inorganic substances which are added to barbeque briquettes to prolong cooking time. They slow down combustion time and retard the heat release of the fuel.

Gas generator or producer gas charcoal: Mainly used in gas engines for the generation of electricity or as an alternative fuel for gas-fired cars.

1.4.2 Pyrolysis Oil

The oil varies very much with the type of raw material. It contains more than one hundred different substances, which once made it a valuable feedstock for the chemical industry. Its growing importance for developing countries stems from the fact that it is a feasible substitute for industrial fuel oil.

Viscosity: This can be controlled by the addition of water, and is measured according to normal standards.

Acidity: Measurement is by analytical titration, without relating the acidity value to particular acids.

Organic fatty acids: These are regular constituents of pyrolysis oil. The most important acids are acetic acid, butyric acid and propionic acid.

Pyrolytic alcohols: Mainly methanol as a regular ingredient of pyrolysis oil.

Calorific value: This is related to the raw materials. Conifers yield oils with high heating values.

Flash point: Determines the ignition properties and is an essential criterion for all boiler fuels.

Flame temperature: Measured under stoichiometric conditions (in an environment of balanced oxygen).

Corrosivity: Pyrolysis oil is aggressive to mild steel. Stainless steel, copper, ceramics, plastics and wood are resistant.

Ashes: These are incombustible particles which have entered into the pyrolysis oil. Reasons: poor distillation equipment, contamination during storage.

Solidifying point: The temperature at which the residues of pyrolysis oil lose their flow characteristics.

Pour point: The temperature at which the residues of pyrolysis oil start flowing.

1.4.3 Process Gas

This is the gas which leaves the retort or converter; it is therefore frequently called retort or converter gas.

A distinction is made between two types of off-gas which have different commercial values:

(a) Off-gas that is not passed through a scrubber/chiller system. It contains the pyrolysis oil vapours and is usually referred to as "high Btu converter gas", meaning that it has a high calorific value.

(b) The residual gas which has passed through a scrubber/chiller system and has been stripped of the pyrolysis oil vapours. It is frequently referred to as "low Btu converter gas", meaning that it is a heating gas of poor calorific value. It is composed of CO, CO_2 hydrocarbons, nitrogen and water vapour.

References

(1)   Plinius, Historia Naturalis, Lib. 11, de pice (23 - 79)
(2)   Glauber (1604 - 1670), Miraculum Mundi, 1653
(3)   C.F. Nordenschoeld, Abhandlungen der Koeniglichen Schwedischen Akademie der Wissenschaften, Jahrgang 1766
(4)   Karl Freiherr von Reichenbach (1788 - 1869)
(5)   FAO Forestry paper No. 41, Rome 1983
(6)   U.N. Conference on New and Renewable Sources of Energy Preparatory Committee, 1981. Report of the Technical Panel on Fuel Wood and Charcoal on its Second Session.
(7)   M. Klar, Technologie der Holzverkohlung, 1910
(8)   W. Emrich, Recycling in Developing Countries, 1982 (210 - 214)
(9)   CARBON INTERNATIONAL, LTD., Neu-Isenburg, FRG, company publication, 1982
(10)   Encyclopedia Britannica, Vol. 5, 1951
(11)   Encyclopedia Americana, International Edition, 1964.

Chapter 2
TRADITIONAL METHODS OF THE SMALLHOLDER CHARCOAL-MAKER

Traditional carcoal-making, which has a long history, is characterized by the following main features:

-   Zero or low investment cost;

-   Use of construction materials which are at hand on the site or available nearby, e.g. clay, soft-burnt bricks;

-   Zero or low maintenance costs achieved by avoiding metal parts in the kiln construction as far as possible;

-   Manpower is not a major concern;

-   The normal raw material consists essentially of wood logs and coconut shells; other types of biomass may be carbonized also;

-   By-product recovery is limited owing to the fact that no sophisticated equipment is employed;

-   It is typically a family or cooperative business.

Much of the charcoal in the world has been made by families or small businesses run on the above lines, using simple technology and low capital investment. Although the technology is simple, it is nevertheless precise and skilful, as will be seen in the following sections. Modern industrial processes are discussed separately in Chapter 3.

2.1 Charcoal Pits and Earthmound Kilns

To save transportation costs, the small-scale charcoal-maker produces his char at the place where he collects the raw material. Because of his frequent movements from site to site, he cannot employ heavy equipment.

Both the pit method and the earthmound kiln offer appropriate solutions. Both require skill, patience and readiness to observe correct working methods at all times and in all weathers. Therefore, the business of charcoal-making with all its "secrets" is usually handed down from father to son and is well guarded by the family.

An important part of charcoal-making experience concerns the insulation of the charcoal pit or earthmound and the control of air flow. If not properly controlled, excess air will cause the charcoal to burn away to ashes and destroy the result of several days' work in a few hours.

The use of earth to keep out oxygen and to insulate the carbonizing wood against excessive heat loss surely goes back to the dawn of history. It is, therfore, worth-while to study its advantages and disadvantages. Obviously this method has survived because of its low cost. Wherever trees grow, earth must be available and mankind naturally turned to this cheap, readily available non-combustible material as a sealant to enclose the carbonizing wood.

There are two different ways to use an earth barrier in the charring process: one is to dig out a pit, put in firewood and cover the whole with excavated earth to seal and insulate the chamber. The other is to cover a pile of wood on the ground with earth, sand and leaves. This cover forms the necessary gas-tight layer behind which charcoaling can take place. Both techniques, if properly carried out, can produce good charcoal within certain limitations. But these methods are wasteful of resources.

The pit and the earthmound methods have been studied by many researchers in the last century (1). According to their results the fixed carbon content (C_{fix}) of the charcoal produced in those kilns varies between 65 % and 80 %, the yields rarely exceeding 15 % based on the weight of the dry fuelwood charge.

Statements that the recovery rate of a charcoal pit or an earthmound operation is "well over 20 %" must be regarded as unfounded. Because of the fact that a significant depletion of wood resources all over the world can be observed as a consequence of wasteful charcoal conversion and the rigid arguing of their defenders, it seems quite appropriate to shed some light on the complexity of the whole problem.

Charcoal conversion efficiency can be defined in terms of either weight:

$$\eta_w = \frac{\text{charcoal output (kg)}}{\text{wood input (kg)}}$$

or energy:

$$\eta_e = \frac{\text{charcoal output (MJ)}}{\text{wood input (MJ)}}.$$

While η_e can easily be calculated if the energy content of wood and its char yield is known, the weight-based efficiency criterion η_w varies depending on just how one defines charcoal. Unfortunately, there is no universally accepted definition of charcoal. However, most of the literature on carbonization agrees that - measured on a dry weight basis - charcoal should exhibit a fixed carbon content of at least 75 % (the other principal components are hydrogen: 4.1 %; oxygen: 15.2 %; nitrogen: 0.8 %; and minerals: 3.4 %).

On the input side, the biomass source wood is composed of cellulose, lignin, hemicellulose, extractives and minerals. In terms of chemical composition, one can distinguish between hardwood and softwood: on the average, hardwood contains about 43 % cellulose, 23 % lignin and

35 % hemicellulose, whereas softwood usually contains 43 %
cellulose, 29 % lignin and 28 % hemicellulose. These
differences in composition affect heat content; the higher
the lignin and extractives content, the higher is the gross
calorific value of wood.

In the process of pyrolysis, lignin promotes char
formation and holocellulose (cellulose and hemicellulose)
promotes the release of volatiles. The heating value of the
primary end prodct - i.e. the charcoal - is determined by
its carbon content. The general equation describing the
relationship between carbon content (C) and higher heating
value (HHV, dry basis) of combustible fuels can be
approximated by the following equation (2).

$$HHV = 0.437 \times C - 0.306 \ (MJ/kg)$$

wich is illustrated in Figure 5.

Another relevant factor here is the species- and
age-dependent moisture content of wood, which directly
affects the net heating value, the ignition properties and
the efficiency of fuel utilization. Fresh wood may have a
moisture content of as high as 67 %, which is also the
practical limit of combustibility. On the average, hardwood
contains 30.2 % moisture, and softwood about 46 %. The
following formula (Tillmann, 1982) can be used to calculate
the net heating value of wood as a function of its moisture
content:
$$NHV \ (MJ/kg) = HHV - 0.2 \ 33 \times MC$$

where

NHV = net heating value (MJ/kg)
HHV = higher heating value (MJ/kg)
MC = moisture content.

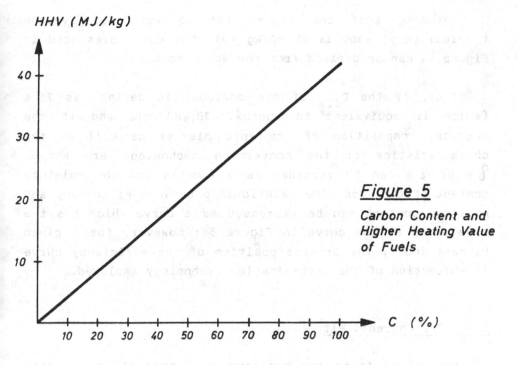

HHV (MJ/kg)

C (%)

Figure 5

Carbon Content and Higher Heating Value of Fuels

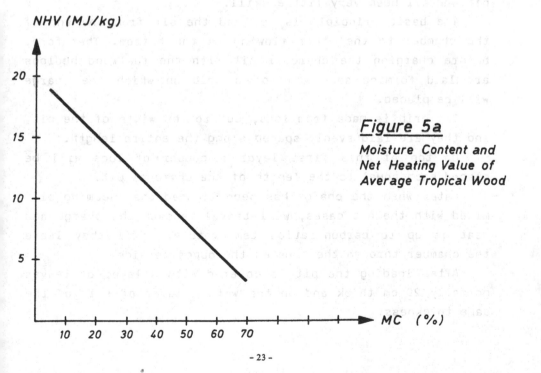

NHV (MJ/kg)

MC (%)

Figure 5a

Moisture Content and Net Heating Value of Average Tropical Wood

Assuming that the higher heating value of average tropical (dry) wood is 20 MJ/kg (3), the curve presented in Figure 5a can be derived from the above formula.

Thus, if the C_{fix} of dry charcoal is defined as 75 % (which is equivalent to approx. 30 MJ/kg), and if the chemical composition of the wood biomass as well as the characteristics of the conversion technology are known, ηw or ηe can be regarded as a function of the moisture content of the wood. The relationship between efficiency and moisture content can be expressed as a curve which has the same shape as the curve in Figure 5a; however, for a given biomass input, the precise position of the efficiency curve is a function of the carbonization technology employed.

2.1.1 The Charcoal Pit

Virtually, there are numerous ways to build a charcoal pit and all need very little skill.

The basic principle is to lead the air from one end of the chamber to the other flowing on the bottom. Therefore, before charging the charcoal pit with the fuelwood bedlogs are laid forming some kind of a crib on which the charge will be placed.

The crib is made from logs, cut to the width of the pit, and they are laid evenly spaced along the entire length.

On top of this first layer a number of logs will be placed each equal to the length of the charcoal pit.

Later when the charge has been lighted the incoming air, mixed with the hot gases, will travel beneath the charge and heat it up to carbonization temperature, until they leave the chamber through the flue on the opposite side.

After loading the pit is covered with a layer of leaves normally 20 cm thick and on top with a layer of soil of the same thickness.

In a typical pit operation burning takes place progressively
from one end to the other.

The produced charcoal of larger pits is not very
uniform, because the burn is difficult to control. Smaller
pits are more efficient, because they have a better airflow
(4).

The large charcoal pit normally takes 25 to 30 m^3 of
fuelwood per burn and the wood is cut in length of 2,40 m.
Miniature pits with a capacity of 4 to 5 m^3 are frequently
in use. Their conversion efficiency is very poor and the
charcoal produced of inferior quality. In Figure 6 the
cross-section of a larger charcoal pit is shown with the
crib structure.

Because it is very difficult to control the airflow in a
pit and the internal temperatures a lot of the fuelwood
charge is burnt to ashes. Another portion remains only half
burnt. Because of false circulation of the gases it was
never dried out and properly heated.

A further problem is connected with falling rain. The
pyroligneous gases tend to condense in the foliage layer and
in the earth used as outer cover. Therefore when rain falls
those condensates are washed back and down to be reabsorbed
by the charcoal (5). They will cause later damage to
jutebags by rotting them and the charcoal when used in
housholds releases unpleasant smoke.

Pit sites should be selected where a deep stratum of
soil can be expected. Where the soil is loose the vents of
the pit must be timbered.

In many cases and for large pits steel sheets are used
for cover before the soil is put on top, thus preventing
contamination of the charcoal by falling soil when the kiln
is uncovered at the end of the carbonization.
According to the FAO Forestry paper No. 41 (4) a team of
five men can produce from a pit size of 6 m x 2.70 m x
1.20 m (depth at the igniting point) and 2.40 m depth at
opposite side 360 tons of charcoal per year.

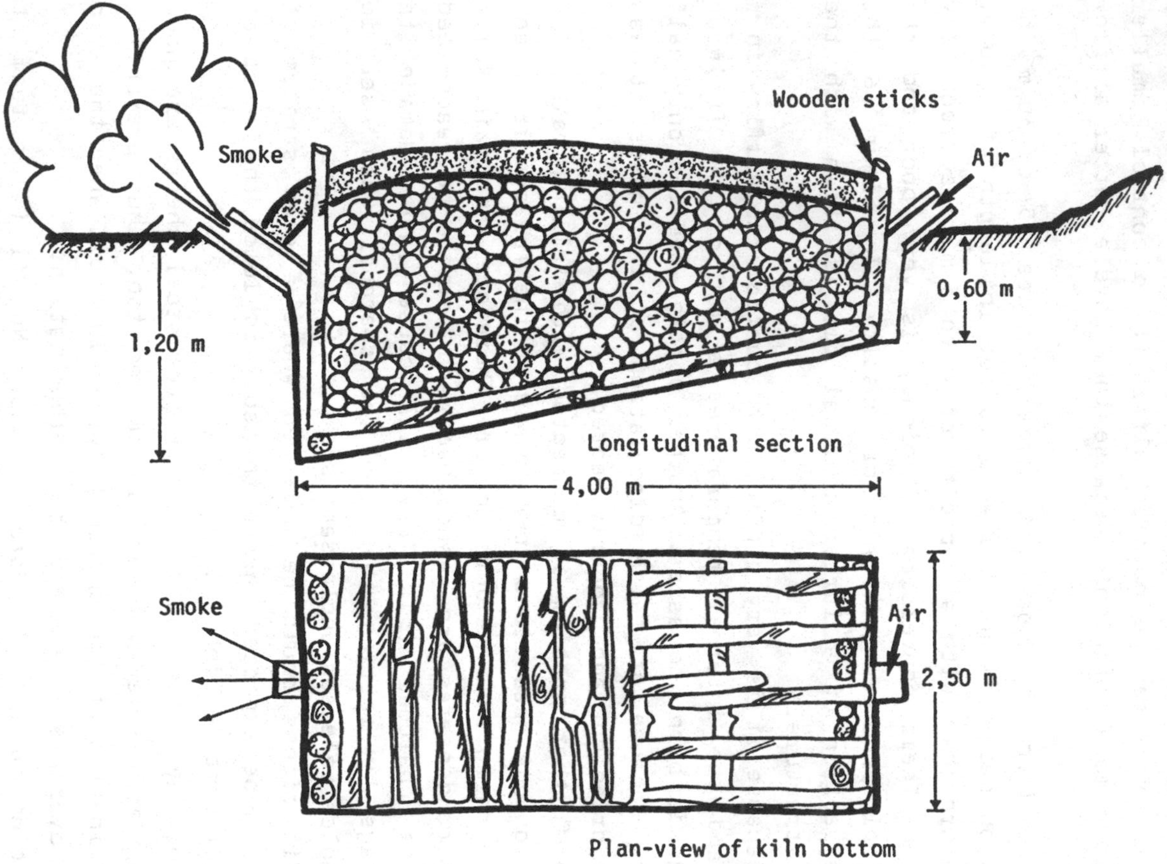

Air

Wooden sticks

0,60 m

Longitudinal section

4,00 m

Smoke

1,20 m

Air

2,50 m

Plan-view of kiln bottom

Smoke

Figure 6. A charcoal pit.

- 26 -

This would have resulted in USD 70.00 per ton in 1983. The indicated cost is for charcoal at the kiln site ready for transportation. The calculation, however, does not allow for labour overhead and profits.

2.1.2 The Earthmound Kiln

The typical smallholder charcoal maker builds a kiln of about 2m in diameter at the base and approximately 1.5 m high as shown in Figure 7. Approximately six to ten air inlets are installed at the base. The smoke and the gases developing during the carbonization are drawn-off at the top through an opening of 20 cm in diameter.

This describes roughly a round earthmound kiln. However, one can find other shaped kilns like rectangular mounds in use as well.

For the preparation of the kiln site sufficient space has to be cleared, levelled and if necessary compacted also. In the case of rectangular or square shaped mounds it becomes necessary to erect posts, several on each side, which give stability to the wood pile and provide a support for the operator when he is covering the kiln with leaves, soil or metal sheets.

Again, as with the charcoal pit, it is most important to provide a good air and gas flow within the mound after lighting. This will be achieved by forming a grid of crossed small logs (maximum diameter 10 cm) which are first laid out and arranged on the ground. On this platform the fuelwood will be stacked.

For the stacking of the wood exist no strict rules and the piles are set up very differently. In general one can say that in spheric earthmound kilns the charcoal maker arranges it vertically and in other mounds horizontally.

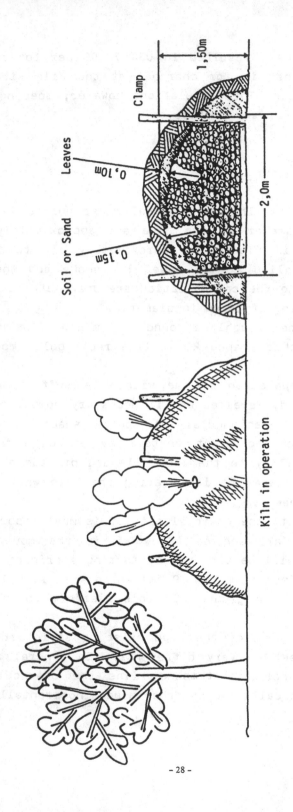

Figure 7. A small earthmound kiln.

- 28 -

All gaps between the logs or branches are filled in with smaller woodlogs, off-cuts, etc. to make the pile as compact as possible which facilitates the direct heat transfer when the carbonization has started.

Also special care has to be taken to the surface of the pile that it shows an even profile and makes a good support for the cover of leaves and soil.

The lower layer of the cover is made with leaves, straw or grass. On top a sandy soil or loam are applied approximately 20 to 25 cm thick.

An important improvement can be achieved by putting in metal sheets before the outer soil cover is made. This will reduce defilement of the charcoal by dropping of soil down during the carbonization, when the mound is slowly sinking, and when it is uncovered at the end.

These metal sheets can be gathered from scrap or cut out of spent oil drums. Depending on the usage they will, however, corrode in a more or less short time and have to be replaced then. Recent calculations have shown that this type of operation may become quite costly in countries where scrap prices are high.

The igniting of a kiln is done either through the air inlet holes on the base or through the centre hole at the top. In every case, to enhance the firing kindling wood, oil soaked fabrics are used or a shovelful of glowing charcoal is put into the centre hole on the kiln top.

After the kiln has "caught fire" the operator observes carefully the colour of the smoke exiting the mound. Dense white smoke will be issued for the first day or days indicating that the water of the fuelwood is being evaporated.

After this initial period, the length of the time depends on the size of the kiln, moisture content of the charge, thickness of the woodlogs, etc., the smoke will turn blue and becomes clear eventually.

PHOTO 1   Small Earthmound kiln in Ghana one hour
          after lighting. (Photo W. Emrich)

PHOTO 2   Discharging charcoal from same kiln
          two days later. (Photo W. Emrich)

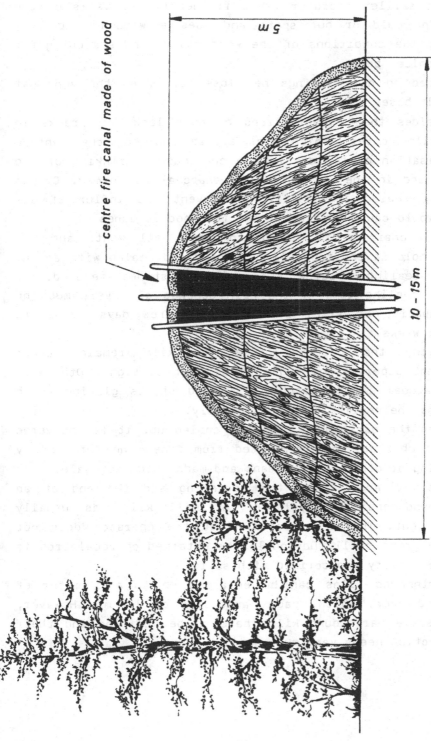

centre fire canal made of wood

5 m

10 - 15 m

Figure 8 A large earthmound kiln with centre firing.

- 31 -

The skilled operator feels frequently the walls of the kiln for cold or hot spots and uses a wooden stick for probing the conditions of the wood charge and searching for uncarbonized parts.

Based on his findings he closes or opens the vents at the kiln base.

Besides these observations he must look for cracks in the kiln cover which do appear during the entire carbonisation cycle, due to the continuous shrinking of the charge and in the same grade as charcoal is formed. Cracks must be sealed immediately to prevent the influx of air which would cause burning of the fuelwood to ashes.

When charring is judged complete, all vents and the centre hole at the top will be closed and sealed with earth. For the sealing of larger kilns clay and stones are used.

The cooling time of the kiln depends also very much on its capacity or size. It may take a few days or up to several weeks.

During the cooling phase the kiln remains under permanent supervision, and must be kept airtight. Otherwise the charcoal inside catches fire and starts glowing which prolongs the cooling time considerably.

When the earthmound kiln has cooled out it is uncovered and the charcoal crop separated from fines and "brands" by screening or other simple means and made ready for sale.

Photo 1 and 2 show the beginning and the end of an earthmound charring cycle of a small kiln, as usually carried out. They also reveal that the operator does not display great skill. Obviously the obtained charcoal crop is of poor quality and defiled with soil.

Earthmound kilns can be enlarged up to a diameter of 15 m and over, whith space capacities of 150 m^3. However, large scale earthmound kilns have to be modified and their construction needs very much skill.

They require centre firing canals as shown in Figure 7. Some details of such a canal structure can be seen in Figure 8. also, which refers to a special kiln type described in the following section.

2.1.3 The Earthmound Kiln with Chimney

These modified kilns represent the most advanced group of the earthmound family. They have been developed in Europe during the middle of the last century, namely in Sweden and in the western parts of Russia.

Very large amounts of charcoal had been produced with these kilns for the growing iron smelting industry.

The elements used for the kiln construction are basically the same as described in the previous section; the grate formed with woodlogs on the bottom, the fuelwood pile (stacked vertically) and the centre hole for firing. However, in addition a chimney is attached (normally only one) which is connected to the pile by an underground flue (6, 7).

The significant improvements rest with this chimney. Since the diameter of it can be determined according to the oxygen demand the kiln can be precisely controlled with the draft of the chimney, which depends also on the height of it, resulting in a higher yield of better charcoal.

Whereas, the traditional earthmound kiln takes the air in through numerous inlets with undefined cross-sections, which will change during the carbonization cycle also (due to the shrinking movement of the kiln shell), the chimney secures an uniform air influx throughout the whole operation time.

The investment costs of these kilns are naturally higher and, therefore, normally metal scrap is used for the chimney construction. Frequently oil drums are welded together and make a good chimney. However, one has to bear in mind that

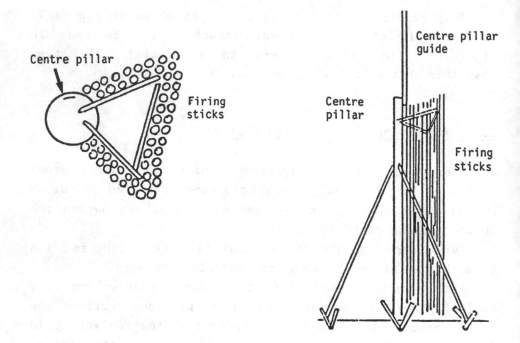

Centre pillar

Firing sticks

Centre pillar guide

Centre pillar

Firing sticks

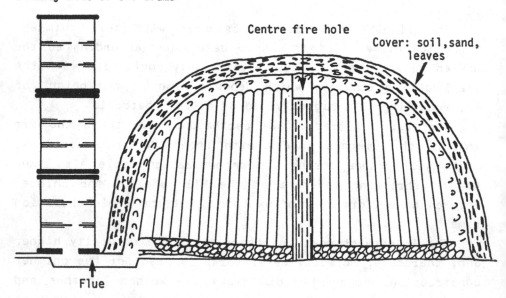

Chimney made of oil drums

Centre fire hole

Cover: soil, sand, leaves

Flue

Figure 9. An earthmound kiln with chimney.

- 34 -

the opening of the smoke stack may have to be reduced and adapted to the size of the wood pile.

In many cases a stack made from firebricks is employed thus eliminating corrosion problems. These chimneys are preferred in charcoal operations with large plant sites and where the fuelwood is permanently carried to.

Figure 9 shows a medium size earthmound kiln with an attached chimney assembled from spent oil drums. The centre pillar as shown in Figure 9 guarantees the firing of the wood pile from top to bottom at the same time. The carbonization will progress from the centre to the edges almost equally and within the entire height of the kiln. Chimney kilns are usually lighted with a torch. During the operation the same observations and probings are necessary as already mentioned in the section dealing with regular earthmound kilns.

Owing to the higher temperatures which can be achieved the charcoal produced in these kilns has a low content of volatile matter and consequently a higher heating value as compared to normal earthmound charcoal.

One disadvantage which is connected with all kilns having a loose shell has to be seen in the always present danger of showing cracks during the carbonization and cooling phase.

This contributed to the fact that many charcoal producers turned to kilns with fixed shells, which are represented by the hangar and round grill types and which will be described later.

2.1.4 The Earthmound Kiln with Tar Recovery

It is only a small step from the chimney-kiln to the charcoal operation with tar recovery. A few but important modifications within in the chimney section are necessary.

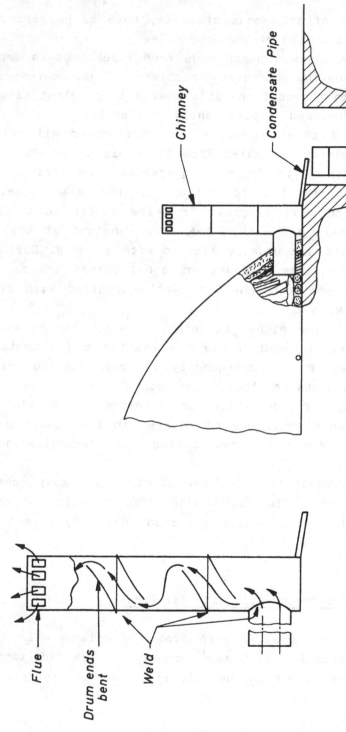

Figure 10 An earthmound kiln with pyrolysis oil recovery.
(_From: FAO Forestry Paper No. 41_)

In the first circumstance the chimney is designed to expel all gases generated during the carbonization process into the ambient atmosphere. In the latter case the chimney serves two purposes: condensing part of the gases and guiding the uncondensable portions to the outside.

Since the investment cost would not allow the installation of chiller and scrubber systems the earthmound operator has to limit himself to the collection of only part of the charcoal by-products. In this case it will be mainly wood tar.

From 100 steres of wood normally 25 tons of condensate can be recovered (8). It consists of water, acids and tar. In practice about 2 tons of tar can be recovered from it (9).

Since all wood tars contain a high proportion of heavy organic chemicals with higher melting and boiling points the condensation of them by aircoolers is quite efficient.

For this reason, the way the hot gases travel through the stack, has to be extended to allow them to dispose of part of their tangible heat within the chimney area, resulting in partial condensation.

This can be accomplished by putting in the chimney some metal sheets which force the gases to flow around them and do not permit them to escape on a direct way.

Figure 10 displays the simple design for a three sectional kiln stack made from oil drums.

In many cases a higher chimney will be required to provide an adequate travel route for the gases.

The condensed tar and oils will be collected at the bottom of the stack and lead through a pipe into the storage drum.

The structure of the wood pile with supporting grate beneath and the shell are built in the same manner as with regular kilns.

Of course, some special structures have been in use with more or less significant effects. The best results the author has achieved were with fixed shell kilns of the hangar type and with a modified Brazilian Beehive kiln.

The economics which can be achieved by the recovery of charcoal by-products will depend very much on the use-market prices for them. In 1983, according to a market survey performed by the author in African and some Asian markets the production cost for charcoal in earthmound kilns ranged between USD 48.00 and 62.00 per ton of lump charcoal. Tar sales would have reduced the production cost by USD 15.00 to 22.00 per ton.

2.2 Charcoal-Making with Portable and Movable Kilns

These kilns are made by modifying oil drums or other containers or are specially designed and built from metal sheets. The latter consist of few sections. They are rapidly assembled or dismantled by the charcoal burner, using simple tools.

Kilns of this kind have been made since the beginning of the century (the Delhemmeau Kiln, 1907) (7), as an easier means than the traditional stack, for the production of charcoal from bulky sawmill waste and the branches and brushwood left after felling. Interest in them quickened when it was found practically and economically feasible to use charcoal gas-producers for propelling motor vehicles. But the use of these kilns reached its maximum in some Western European countries during the 1939 - 45 war when petrol supplies were short and charcoal became by far the most commonly used fuel for combustion engines. The vogue of portable kilns passed with the return of normal fuel supplies to these countries. The kilns are now made by only a few of the many manufacturers of ten years ago.

In this type of kiln, carbonization is discontinuous, being carried out by the combustion of part of the timber to be carbonized. For the most part, these kilns work on the downward, rarely direct, draught principle. Observations (10) made with the help of regularly spaced pyrometers inside kilns of the stack type show that carbonization works

- 38 -

in the same direction as in the traditional pit or
earthmound kiln, e.g. from the top downwards and from the
middle outwards. Carbonization temperature is usually 440°C,
but may be higher and even slightly exceed 600° C during
"flare-ups".

Portable kilns have one great advantage over the
traditional stacks in that they appreciably reduce the time
of the carbonization operation (about 48 hours), chiefly
because the preliminary "dressing" of the stacks with earth
is dispensed with. Moreover, the carbonization process
requires much less surveillance, so that there is some
economy in manpower. Furthermore, the charcoal produced is
free from earth and gravel and this is important where it is
to be used for gas-producers or combustion engines.

The chief criticism directed at these kilns emphasizes
that they do not facilitate the collection of by-products,
and that their charcoal yield is low compared with the
weight of wood carbonized (4). The yield, however, can be
appreciably increased if the various operations are carried
out carefully, especially charging and control of the air
intake.

2.2.1 The Tongan Oil Drum Kiln

The Tongan oil drum kiln is a low-capital, small-scale
method of producing charcoal from wood (11).

As the name suggests, this kiln was developed in Tonga
when the need arose for a low-cost method for converting
coconut shells and wood to a salable product. The solution
was a simple-to-contruct and use charcoal kiln made from a
200 litre drum. Other drum kilns have been made in the past,
but because of their design have not been particularly
efficient or easy to use. The Tongan kiln, however, takes
advantage of the drum shape to produce good quality charcoal
simply.

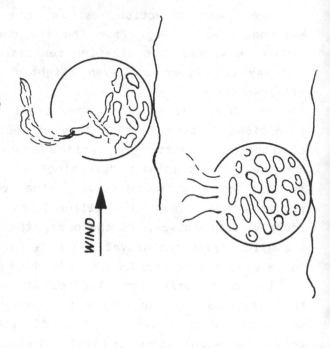

WIND

WIND

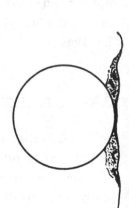

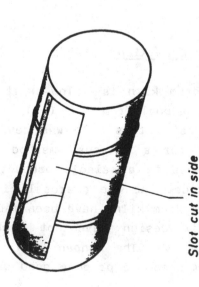

Figure 11   The Tongan oil drum kiln

200 litre oil drum;
both ends sealed

Slot cut in side
approx. 20 cm wide

The key to the Tongan design is the opening, about 20 cm wide, which is cut along the vertical side of the drum (see Figure 11). With the drum lying on its side and the opening facing the prevailing breeze, a fire is built inside. The drum is rolled so that the opening reaches a progressively higher position. This allows more and more firewood to be added until the drum is completely full. It is important that a good fire is kept burning without being smothered by the addition of new wood. As the fire burns down, more wood is added - the whole process taking up to an hour. When no more wood can be added and the fire is burning strongly, the drum is ready to be sealed. (The charring process has already begun in the lower part of the drum where the air cannot reach). The drum is rolled over so that the opening is facing the ground. Clay or sand is packed around the bottom so that no smoke can escape. At this time, the entire drum is sealed and can be left for 6-8 hours to cool. When it is completely cool, the drum can be rolled over again and the charcoal removed (see Figure 11).

The Tongan kiln possesses several advantages:

- Low cost: The only costs involved are the drum and the tools to cut it open (hammer, chisel, bush knife).

- Transportable: One person can carry it to the site of the firewood.

- Simple to use: There is less danger of being burned than with some larger kilns and retorts because the major movement is by rolling and the small size makes the fire manageable. It does not require highly skilled labour.

- Efficient: Tests have shown that the output of charcoal per kilogram of firewood equals that of the larger retorts.

- It can deal with small amounts of firewood. One needs merely to increase the number of drums if greater quantities are to be processed.

The disadvantages are:

- Firewood must be cut to fit the drum.

- Drums last at most 6 months.

- Charcoal in contact with the ground during the cooling part of the process may absorb moisture. To prevent this, the original section of the drum that was removed to make the opening can be replaced as a door using a wire hinge.

2.2.2 The Philippines Kiln

This kiln was first designed and used in the Philippines, in particular for the carbonization of coconut shells. It employs an oil drum set up vertically (12).

The Philippines kiln is made from an old oil drum in which the top and bottom lids are still firmly in place. Reject oil drums can usually be bought cheaply at oil company depots, and these are quite suitable. In addition to the drum, a flat circular sheet of thin metal, about 50-54 cm in diameter, with a central hole 10 cm in diameter, is required. This forms a movable lid for the kiln. For convenience, a handle can be bolted or tack-welded onto this lid. A smaller piece of flat metal 15 cm in diameter is used to form a cap for the central hole in the lid when charcoaling is completed.

The kiln is constructed as follows. At the top of the drum there are two holes. The plugs should be removed from these holes and, using an oxy-acetylene welding torch, the

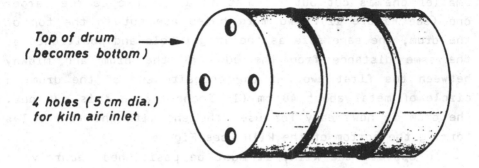

Top of drum
(becomes bottom)

4 holes (5 cm dia.)
for kiln air inlet

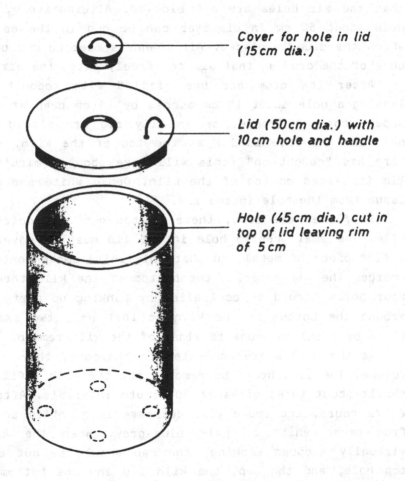

Cover for hole in lid
(15 cm dia.)

Lid (50 cm dia.) with
10 cm hole and handle

Hole (45 cm dia.) cut in
top of lid leaving rim
of 5 cm

Figure 12   The Philippines oil drum kiln

smaller one is cut out so that it is as big as the larger one (see Figure 12). Two more holes are cut in the top of the drum, the same size as the larger hole and positioned at the same distance from the edge of the drum and midway between the first two. At the opposite end of the drum, a circle of metal about 48 cm (16 inches) across is cut out. The drum is now ready for use, the end with the four holes forming the bottom of the kiln (see Figure 12).

To operate the kiln, it must be positioned securely on bricks or pieces of metal pipe or rail about 5 cm high, so that the air holes are not blocked. Alternatively, a round hole about 50 cm in diameter can be dug in the earth over which the drum is placed. Air channels must be dug under the edge of the drum so that air can freely enter the air holes.

After the drum has been filled with coconut shells, leaving a hole about 15 cm across by 15 cm deep at the top, paper, coconut fronds, or an oily rag are placed in this hole and the fire is lit at the top of the kiln. When the fire has "caught on" (this will take about 5 minutes), the lid is placed on top of the kiln. Dense white smoke should issue from the hole in the lid.

If flames are seen, there is too much air entering the kiln. The small centre hole in the lid must be covered with a flat piece of metal, so that only white smoke continues to emerge. The air entering the bottom of the kiln through the four holes should be controlled by banking up earth or sand around the bottom of the kiln so that only two small gaps (5 cm by 1 cm) on opposite sides of the kiln remain.

As the shells are converted to charcoal, they reduce in volume. The lid should be removed and the kiln refilled with shells about three times at 30-minute intervals. After about 2 1/2 hours, the smoke will decrease in quantity and change from dense white to pale blue-grey. When the kiln has virtually stopped smoking, the cap should be put over the top hole, and the cap, the kiln lid and the bottom of the kiln must be sealed securely with sand or earth so that no air can enter the kiln. Any air entering the kiln will reduce the charcoal yield.

Four hours after sealing, the kiln should be cool to the touch and ready for discharge. Each kiln should yield 12-25 kg of charcoal per firing. A single worker should be able to operate up to 10 kilns on a daily cycle.

2.2.3   The Black Rock Forest Kiln

This kiln is in one piece, the part being electrically welded to form a bell of steel sheeting. The dimensions are shown in Figure 13. Capacity is about 2.8 m^3.

The lower part is pierced with holes, 4 for the reception of steel smoke outlets, the others acting as air intakes. The upper part has a hole fitted with a metal cap.

The kiln was designed with a view to its use as part of a fixed charcoal burning yard, consisting of four kilns which can be hoisted on to eight cement emplacements. The latter are distributed in a circle, in the centre of which is a derrick for moving the kilns (see Figure 14, 15).

Large stocks of wood are piled near the emplacements, and while carbonization is taking place on emplacement A, with kiln no. 1, (Figure 15), the crew prepares the stack on emplacement B.   When carbonization is completed on emplacement A, number one kiln is lifted therefrom and placed over the stack already prepared on emplacement B. Firing takes place at once, and while carbonization is in progress the crew sacks the charcoal made on emplacement A and prepares the fresh stack on A.

When carbonization is completed on emplacement B, the kiln is hoisted and placed over the stacked wood on A, and so on. It can be seen that when this process is repeated for each of the emplacements A, B, C, D, E, F, G, H, and for each of the four kilns, there are no slack periods in the crew's working time.

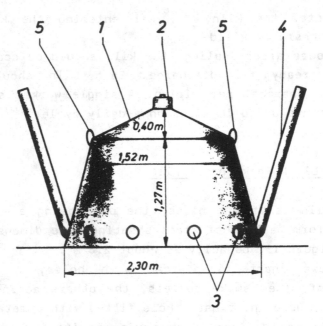

0,40 m

1,52 m

1,27 m

2,30 m

1 = Steel sheet
2 = Lid for ignition hole
3 = Air inlets at the bottom
4 = Chimney
5 = Hoisting rings

Figure 13   The Black Rock Forest kiln.

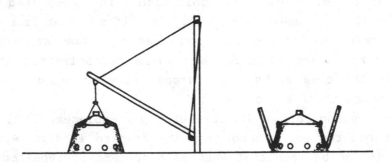

Figure 14   Handling the kiln with a derrick.

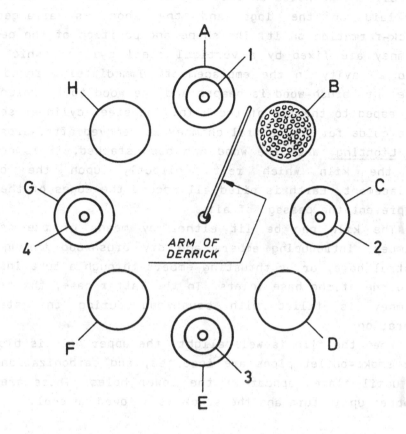

A - H = Concrete emplacements for kilns
1 - 4 = Kilns in operation

Figure 15 *A battery of four Black Rock Forest kilns in operation*

Preparation of the wood. Logs are laid radially on the emplacement, leaving channels between them to correspond to the air intake holes in the base of the kiln. The wood grate is laid on the logs and the wood is arranged in stack-formation on it. The shape and position of the central chimney are fixed by a vertical steel cylinder which fits into a cavity in the emplacement. Immediately around this pipe, dry brush-wood is heaped and the wood to be carbonized is heaped to the top of the kiln. The steel cylinder serving as a guide for the central chimney is removed afterwards.

Lighting. When the wood has been stacked, it is covered by the kiln which rests directly upon the cement emplacement. Earth is piled all around the edges of the kiln to prevent the passage of air.

The kiln can be lit either by means of the central chimney, introducing embers and dry brushwood through the central hole, or by thrusting embers through a tube inserted into one of the base inlets. In the latter case, the central chimney is filled with brushwood during the stacking operation.

When the fire is well alight, the upper hole is blocked, the smoke-outlet pipes are inserted, and carbonization goes on until flames appear at the lower holes. These are then stopped up in turn and the stack is allowed to cool.

2.2.4 Sectional Metal Kilns

Several versions of these kiln types have been on the markets and in use. Especially during war times the army employed these kilns to obtain a clean charcoal for their own supply.

They all consist of at least one cylindrical section with a conical cover and two or more smoke stacks. On the kiln base are usually four air channels fitted and the cover is provided with four equally spaced steam release ports.

The kilns are made from sheet metal and can be built by local craftsmen in a workshop which has basic welding, rolling, drilling and cutting equipment.

Figures 16 and 17 show a sectional kiln and the way the wood grate on the bottom of the kiln is arranged.

The principles which apply for the charging and operation of the earthmound kilns described in the previous sections are almost the same for the sectional metal kilns.

The popularity and the widespread use these kilns once enjoyed is mainly due to the fact that their sections can be separated and carried to another plant site.

However, several facts have hampered their promotion during the past five to ten years gradually:
- rising iron and steel prices have made the kiln unaffordable for many charcoal makers especially in developing countries
- the kiln shell, if not properly designed and contructed, can be deformed during operation and the repair requires machinery which would be rarely available in a remote charcoal camp
- the kiln tends to corrode and the lifespan is short in comparison with firebrick kilns.

The Tropical Products Institute (TPI), a scientific unit of the Overseas Development Organisation of the U.K. has gained considerable experience in operating sectional metal kilns of various design. The institute has evolved a kiln with good durability which is considered to be optimal in economy also.

The institute recommends a two sectional kiln, consisting of two interlocking cylinders with a conical cover. The kiln is supported on eight air inlet/outlet channels, arranged radially around the base (13).

During charring four smoke stacks are fitted onto alternate air channels. The cover has four equally spaced steam release ports which may be closed off with plugs.

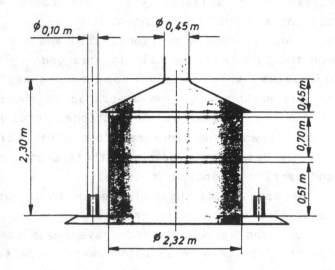

Figure 16   *A portable metal kiln*

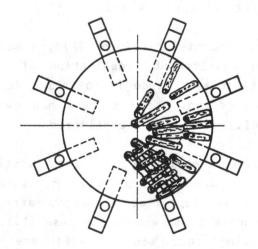

Figure 17   *Air inlet channels at the bottom of a portable metal kiln with wood grate*

The size of the wood to be charged to the kiln must not exceed 60 cm in length and the maximum diameter is 20 cm. Wood with a diameter greater than this should be split before kilning. The TPI kiln takes approximately 7 m^3 of fuelwood.

After placing the lower section of the kiln onto the supporting air inlet channels the bottom of the kiln is laid out with stringers forming a grate. Care has to be taken, that the air channels protruding into the kiln are not blocked by the wood charge.

The fuelwood is normally placed horizontally in successive layers, filling in as many gaps as possible. It is advisable to place the thicker logs in the centre of the kiln where they will be more exposed to the heat as close to a wall.

For the lighting of the kiln several points at the base will be prepared with kindling material and for the ignition a flame is necessary.

During the carbonization colour of the smoke, wall temperature and leaks have to be watched. For the cooling of the kiln all openings are closed and the smoke stacks were taken down before.

The author has experimented with various kilns of the TPI type in East Africa and has established the following working schedule with hardwood (acacia d.):

Two experienced operators

1. day	Loading time	2 hours	08:00 - 10:00
	lighting and draft reducing	1 hour	10:00 - 13:00
	Charring time	19 hours	
2. day	Unloading	2 hours	08:00 - 10:00

For the operation the following tools were in use:
one crosscut saw, one axe, two wedges, one sledge hammer,
one cutlass, one rake.

According to the TPI the following materials and their
quantities are necessary for the construction of the kiln:

Part	Material	Quantity
Base section		Sufficient to cut:
Top and bottom rings	50 mm x 50 mm x 3 mm mild steel (m.s.) angle	6 pieces, each 2 430 mm long
Body	3 mm m.s. sheet	3 pieces, each 2 430 mm x900 mm
or:		
Body	3 mm m.s. sheet	3 pieces each 2 480 mm x 900 mm
Upper section		
Top ring	50 mm x 50 mm x 3 mm m.s. angle	3 pieces, each 2 398 mm long
Bottom ring	50 mm x 50 mm x 3 mm m.s. strip	3 pieces, each 2 398 mm long
Body	2 mm m.s. sheet	3 pieces, each 2 448 mm x 900 mm

Top cover

Cover sectors	2 mm m.s. sheet	2 pieces,
Steam ports	50 mm x 3 mm m.s. strip	4 pieces, each 630 mm long
Lifting handles	10 mm diameter m.s. rod (concrete rein- forcing bar)	4 pieces, each 500 mm long

Steam port covers (4 per kiln)

Bodies	Either 50 mm x 3 mm m.s. strip OR 140 mm diameter steel pipe. (Use pipe if available)	4 pieces, each 440 mm long OR 4 rings, each 50 mm wide
Top discs	3 mm m.s. sheet	4 discs, each 190 mm diameter.
Handles	5 mm diameter steel rod (concrete rein- forcing bar)	4 pieces, each 180 mm long

Base channels (8 per kiln)

Channel sections	3 mm m.s. sheet	8 pieces, each 500 mm x 500 mm
Spigots	Either 3 mm m.s. sheet	8 pieces, each 375 x 150 mm

```
                              OR              OR
                              120 mm diameter  8 pieces,
                              steel pipe       each 150 mm
                                               long
```

Smoke stacks (4 per kiln)

```
                              Thin-walled steel 4 pieces,
                              pipe              each 2 300 mm
                                                long
```

This list may be used by the prospective charcoal-maker
as guide to determine his own investment costs according to
the domestic metal prices and to find out about the
maintenance costs of local shops.

For the filling of sacks the use of a sieve chute is
very practical. The chute is positioned in a sloped manner
and should have a width on the upper end of 1.000 mm and on
the lower side 300 mm to keep the sacks wide open.

The screen can be made from square wire mesh and the
mesh sizes may vary between 10 and 40 mm to separate the
charcoal fines. The classification of charcoal for exports,
however, will require sieving by standardised screens.

2.2.5 The Carborion Kiln

Although the manufacturer of this type of kiln ceased a
few years after the Second World War, it is worth mentioning
for its originality. As far as is known, it is the only
portable kiln which produces charcoal in a retort with
external heating, and it may be useful to describe its
principles.

The Carborian Kiln consists of a tank of thick sheet
metal which takes the wood to be carbonized. The bottom of
the tank is perforated and it is fitted with an air-tight
lid. There are models of one-half and one stere capacity
respectively (see Figure 18).

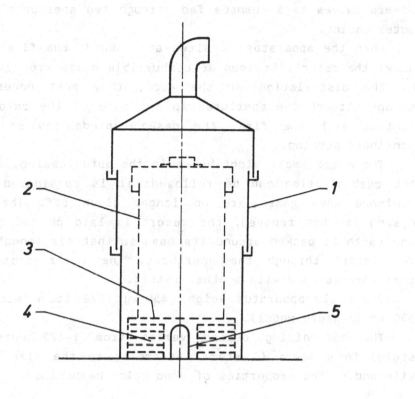

1 = Sections of kiln shell
2 = Retort
3 = Apertures in retort base
4 = Brick bed
5 = Door in outer casing

Figure 18 *The Carborion kiln*

The tank rests on a small layer of bricks, and the whole is enclosed in a sheet metal casing of fitted sections which are easily dismantled. The free space between the brick layers serves as a furnace fed through two apertures in the outer casing.

When the apparatus is sited and loaded, the fire is lit under the retort. As soon as combustible gases are given off by the distillation of the wood, they must necessarily escape through the apertures in the base of the retort. In contact with the fire, the gases explode and so ensure continued burning.

There are small sight-holes in the outer casing, so that the carbonization can be followed: it is considered to be finished when gases are no longer given off. The outer casing is then removed, the retort is laid on the ground, and earth is packed around its base so that air cannot enter the retort through the apertures. The outer casing can therefore be used with another retort.

The whole apparatus weighs 450 kg (1/2 stere model) or 590 kg (1 stere model).

The carbonizing period varies from 1-1/2 hours (1/2 stere) to 4 hours (1 stere) according to the size of the kiln and to the properties of wood being carbonized.

2.3 Concrete and Brick Kilns

In this category one finds the most effective kilns, provided that they are properly constructed and operated. Some types have proved their economic viability over more than fifty years. The number of these kilns in operation at present can hardly be estimated, but it is certainly close to 250,000.

The kiln design is simple, the investment capital requirements are low, and a surprisingly good quality of charcoal is produced both for household fuel and industrial uses.

Both concrete and brick types must comply with a number of important requirements to be succesful. The kiln must be simple to construct, relatively unaffected by thermal stresses on heating and cooling, and strong enough to withstand the mechanical stresses of loading and unloading. It must be unaffected by rain and weather over six to ten years.

The kiln must permit control of the entry of air at all times, and for the cooling phase there has to be a provision for effective hermetic sealing. It must be of reasonably lightweight construction to allow cooling to take place fairly easily and yet provide good thermal insulation for the wood undergoing carbonization, otherwise the origination of cold spots due to wind impact on the kiln walls will prevent proper burning of the charcoal and lead to excessive production of partially carbonized wood (brands) and low yields. The ability of the brick kiln to conserve the heat of carbonization is an important factor in its high conversion efficiency of wood to charcoal.

The design of these kilns has been refined over a long period of time. They can be differentiated by their shape into hangar kilns (rectangular of square shape) and round brick kilns.

- Hangar kilns: Missouri kiln, Ottelinska furnace, Cinder-block kilns.

- Round brick kilns: Argentine kiln (half-orange), Brazilian kiln (beehive), Schwartz furnace.

The Missouri, Cinder-block, Argentine and Brazilian kilns burn part of the charged wood within the kiln to carbonize the remainder. The Ottelinska, Reichenbach and Schwartz furnaces use the hot flue gases from a central fire grate, passed through the kiln to supply heat for drying and heating the wood to start carbonization. The Ottelinska,

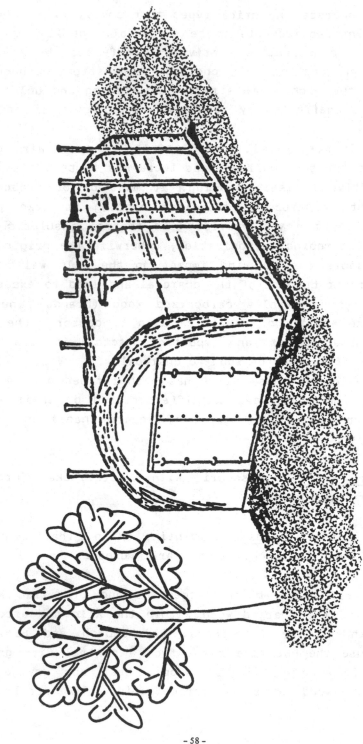

Figure 19. The Missouri charcoal kiln

- 58 -

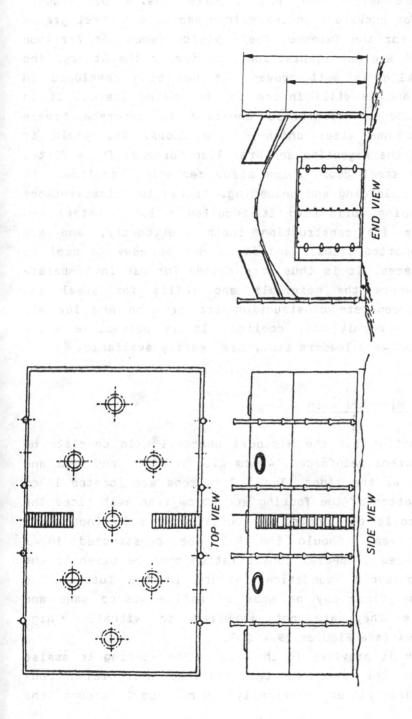

END VIEW

TOP VIEW

SIDE VIEW

Figure 20   Plan and elevation of the Missouri kiln

Schwartz and Reichenbach types require considerable amounts
of steel for buckstays on the kiln chamber, and steel grates
and doors for the furnace. Their yields (when the firewood
is counted) are not in practice superior to the others. The
Missouri kiln is well proven, it has been developed in
practice, and is still in use in the United States. It is
usually made of reinforced concrete or concrete breeze
blocks and has steel chimneys and doors. Its yield is
similar to the Argentine and Brazilian furnace. It is fitted
with large steel doors which allow mechanical equipment to
be used for loading and unloading. It has two disadvantages
for developing world use: it requires a lot of steel and
cement for its construction (both are costly, and are
usually imported items), and it is not so easy to cool as
other furnaces. It is thus more suited for use in temperate
climates where the materials and skills for steel and
reinforced concrete construction are at hand and low air
temperatures permit easy cooling. It is attractive where
labour, front-end loaders etc., are readily available.

2.3.1   The Missouri Kiln

The footing for the Missouri charcoal kiln consists of
concrete, steel reinforced, 45 cm wide in front and rear and
40 cm wide at the sides. Two 1,3 cm rods are located 15 cm
from the bottom of the footing and 15 cm from each side. The
footing should be located in undisturbed soil (not in a
filled-in area). Should the kiln be constructed in a
filled-in area , special consideration must be given to the
footing to avoid foundation failure in the future. The
footing and floor may be made of native washed sand and
gravel, as they are not required to withstand high
temperatures (see Figures 19 - 20).
A notch is provided in the top of the footing to assist
in locking the wall to the footing. Reinforcing rods
(1,3 cm) are placed vertically 30 cm apart around the

foundation where the wall joins the footing. These rods should extend 25 cm into the footing and have a 12 cm leg at a 90 degree angle to ensure a strong bond (14).

The floor is a 10 cm concrete slab with 15 x 15 cm # 10/# 10 steel reinforcing mesh. The floor has a crown down the centre line of the kiln and at the low or drain end of the kiln the floor slopes 15 cm to either side so that the liquors processed from the wood will drain out of the 6 cm pipes built in the footing. Any fill under the floor should be of gravel packed over wetted soil to ensure a firm base.

Undesired air entrances are sealed with clay, mud or ashes. Doors at each end of the kiln provide uninterrupted loading and unloading of the kilns when more than one kiln is being operated.

Construction details (14)

Six 1,50 m length of used boiler flue pipe, 10 cm in diameter, are required for the six air inlet holes in side walls. Four pieces of 4 m long, 10 cm diameter boiler tubes are desirable to pipe air from the doors to near the centre of the kiln during burning. Seven caps are required for the seven fresh air ventilation holes made from truck rims. Eight chimney caps are required to plug the chimneys during the cooling phase. Two door stops are necessary for the holes provided in the footing.

A suggested ladder and catwalk are indicated in the figure. These provide ready access to the top of the kiln. A ladder may be used to reach chimney tops for capping and uncapping.

The volume of this kiln design is 180 m^3. However, it is expected that only 165 m^3 of wood would be loaded into it for burning.

The walls and roof are made of expanded shale aggregate which has the same coefficient of expansion as the Portland cement with air entraining agent. This will minimize cracks

-61-

in the kiln under the temperatures encountered in charcoaling. Some cracks are to be expected due to the different coefficient of expansion of the steel and concrete. The air entraining agents cause minute air bubbles in the concrete, and among other things, make the concrete a better insulator which helps prevent loss of heat during the coaling cycle.

The front and rear walls are 30 cm thick. The side walls are 25 cm thick. This thickness provides heavy duty unloading abuses. These abuses are, for example: throwing the cordwood into the kiln during loading and often hitting the wall, mechanical equipment ramming the wall, an automotive truck bumping into the walls whilst loading the kiln.

Bent plates (19 mm thick) topped together with a 30 cm I-beam on its side form the door facing for kiln protection.

Each side wall contains four 15 cm diameter tiles for chimney entrances. A 12 cm air inlet hole is located half-way between each chimney.

15 mm diameter reinforcing steel rods are located 30 cm apart and centred in the walls and roof. Each intersection is tied together with wire for additional strength.

The roof contains seven ventilation holes for fresh air during unloading of the kiln. Three of these are down the top centre line of the kiln and two are on either side. Their exact location may be varied within limits. To reduce expense, these holes may be made from used truck rims cut in half to make two forms from each rim.

Doors are large enough to provide truck and tractor entrance for loading and unloading. The doors for the kiln are made of 9 mm thick steel plate, 1.50 m wide and 2.50 m high. There are two doors closing at the centre of each end of the kiln. Each door is hung on four 40 cm heavy duty hinges. Fifteen 2 cm bolts are used to secure the door tight on each end of the kiln. These bolts are spaced about 30 cm apart on the top.

Photo 3   A Missouri kiln. The shell is dangerously
           cracked as a result of faulty operation.
           Ghana.(Photo W. Emrich)

Photo 4   Side view of kiln with two smoke pipes
           and air inlet holes at the bottom. Ghana.
           (Photo W. Emrich)

Operation

A crew of two men is needed for loading and unloading equipment with a front-end loader and truck. One operator per shift is sufficient to control the burning and one man per shift can supervise a number of kilns.

The burning of the kiln is controlled in a similar way to the portable metal kiln. The gas circulation system is similar. Yields are usually better because the better thermal insulation and greater ratio of volume to surface area means that the endothermal heat of carbonization is better utilized and the kiln is not so much subject to the cooling effects of winds and rain as the uninsulated metal kiln.

Missouri kilns are usually equipped with thermocouples to read the temperature at several points within the kiln. This is important with such large kilns as it enables cold and hot spots to be readily detected and corrective action taken by the operator by closing or opening air vents along the base of the kiln. The cooling process can also be checked so that the kiln is opened only when the temperature of the charcoal is low enough. This avoids fires which, in such large kilns, are not easy to control even with mechanical handling.

The sudden entry of large amounts of air into a burning kiln may cause an explosion. Numerous minor explosions have been reported in Missouri with no personal injuries. The most serious explosion reported was the partial destruction of a kiln. The top was blown off. The person closest to the kiln was 300 m away.

It is commonly reported that kilns "puff", and lids on air ventilation holes in the top are blown off. Such occurrences can be avoided by proper sealing with clay or soil, preventing the sudden entry of air, as mentioned above.

The photos 3 and 4 exhibit a not so rare damage of the kiln shell caused by failure of the construction material and faulty operation.

The kiln cycle is usually about 25 to 30 days, depending on cooling rates. The capacity of two 180 m^3 kilns is about equal in wood consumption to a standard battery of seven medium-size Brazilian kilns. But because the cycle time is different, the utilization of labour is not so efficient as it could be unless there are more than two kilns to a battery. Utilization of mechanical equipment is not optimized unless the number of kilns is sufficient to keep it working more or less continuously.

The cycle time of a Missouri kiln in a warm climate can be at least one month, made up as follows:

Loading: 2 days 2 men plus machines.
Burning: 6 days 2 men on 12 hr. or 3 men on 8 hr. shift.
Cooling: 20 days (min) 1 man part-time supervision.
Unloading: 2 days 2 men plus machines.

The total time is 30 days. If machines are not available, the cycle time can stretch to two months or more.

The Missouri kiln's greatest advantage compared with brick kilns is the possibility, in fact the necessity, of mechanical loading and unloading.

Its disadvantages are the high cost, due to the use of large quantities of steel and concrete, and its immobility. Unlike brick kilns, it cannot be demolished and rebuilt. Hence a ten-year wood supply must be available within economic haul distance of any group of kilns. The amount of wood for a group of three 180 m^3 kilns would be 60.000 m^3 approximately. About 2.000 ha of forest capable of yielding 30 m^3 per hectare would have to be set aside for ten years to supply this amount of wood. Such an area would give a mean haulage distance of about 1 km which is reasonable.

It is the belief of FAO/Rome that Missouri kilns are not suitable for technology transfer to developing countries because they require large quantities of costly imported items like cement and steel. Also the long cooling time presents a drawback which becomes sensitive in areas with a hot climate.

2.3.2 Cinder-Block Kilns

The masonry-type walls of these kilns should be supported by a continous perimeter-type reinforced concrete footing. The footing should extend at least 25 cm below the surface of well-drained ground. If only intermittent winter operation is planned in areas where the ground freezes, the footing should extend below the frostline. If continous winter operation is planned or if frost seldom or never occurs, a minimum footing depth of 25 cm may be used (see Figures 21 - 22).

It is very important that the first course of blocks be carefully laid in a full bed of mortar. If the base and second courses are accurately laid, the balance of the blocks will go into place with all joints perfectly broken. All blocks are laid with the hollow cores vertical. If hollow core blocks are used in the top course, all cores must be filled with mortar to prevent channelling of air into the kiln through cracks that might develop on the inner wall surface (15).

Neat cement or the standard 1:3 mix of cement and sand may be used to level off the tops of the front wall blocks over which the bottom of the angle-iron lintel must rest in order to seal off the ceiling cover with sand. A piece of bright sheet metal may also be used between the angle lintel and the top of the wall to permit the lintel to slide more easily along the top of the wall. A loose brick may be used to retain the sand at the end of the lintel.

One or two courses of brick are commonly laid around the top edge of the kiln to act as a coping. The purpose of the coping is to help prevent the ceiling sand from being washed and blown off the ceiling steel. Sections of steel rails or I-beams can serve both as a coping material and as a support for the ceiling beams. They also distribute the weight of the ceiling beams along the side walls and restrain the walls from bowing outwards at the top.

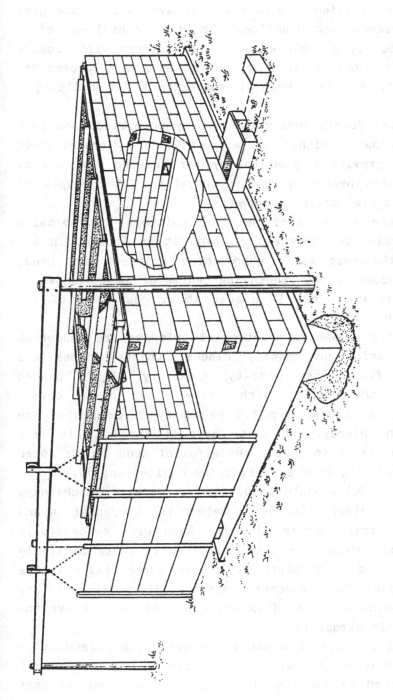

Figure 21 A cinder-block charcoal kiln

(Courtesy of Southeastern Forest Experimental Station,
Ashville, N.C.)

Good workmanship is very important when laying the walls. If an experienced mason is not available, some good manual of recommended practices should be consulted before attempting to lay any block. Such publications will provide dimensions of the various types of blocks, recommended mortar mixes, and many helpful suggestions for building a good wall.

All mortar joints must be carefully compressed and left neat and compact, either in a concave or V-shape. Such joints will provide a good valley for sealing compound as sealing becomes necessary during operation. Other types of mortar joints are not recommended.

Either one or as many as four chimneys are normally used. When only one chimney is used, it is located in the centre of the rear wall. When three chimneys are used, another is added at the lengthwise centre of each side wall. When four are used, one is placed at or near each of the corners of the kiln.

15 or 25 cm diameter sheet-metal chimneys are supported on loose resting on masonry blocks. Chimney bases are constructed from loose pier-type masonry blocks placed directly on the ground, with loose steel plate covers resting on the blocks. Two 1.0 by 40 by 50 cm plates are used at each chimney. One plate next to the kiln is left loose and removable so that a shovelful of sand can be added for closing-off the chimney during the cooling period.

Partial or full-length insulation of metal-type chimneys is helpful in colder climates to retard the condensation and build-up of tars. Chimney tile has been used successfully inside masonry block chimneys, particularly those partially or wholly buried in an earth fill. Some commercial chimneys have been made from concrete masonry block without any lining. All chimneys should extend at least 30 cm above the top of the kiln structure.

Front-entry air supply has given satisfatory results in both one- and three-chimney designs. Front air entry is most easily obtained by hanging the sliding metal door so that there are about 8 cm of clearance at the floor line. It is then a simple matter to seal the opening a litle at a time with earth or sand as less air is required. Some operators prefer to use blocks laid on their sides in front of the door to baffle heavy winds during coaling. When a laid-up masonry door is used, blocks of the first course are laid on their sides so that cores are horizontal. Then earth or sand is used to close off the openings as less air is needed.

Sidewall air ports are made by omitting half blocks in the base course at predetermined locations along the side walls. Each opening can be lined with chimney tile or brick if desired, or used without lining if cinder-concrete is used. Air ports lined with chimney tile can also be used as chimney openings.

Although a roof structure is not a part of the kiln, it is desirable for shedding rain or snow and to protect ceiling parts from early corrosion. When roof trusses are used to support the steel ceiling, the roof structure becomes essential. The kiln may be roofed in any convenient way, with either boards or sheet-metal roofing. Trusses can be assembled from ordinary woodlot lumber. A simple shed-type roof with poles of material slabbed on two sides or squared for rafters and beams would also furnish the necessary ceiling protection.

Operation:

Three types of wood are generally used: cordwood; sawmill slab and edging stock; and blocks and short-length material from sawmills or wood-manufacturing plants. Cordwood and slabs and edgings are usually 1.20 m or more in length, and short-length discarded material may vary from about 8 to 40 cm length.

The manner in which a kiln is charged depends primarily on the types of wood and the location of openings for regulated air entry and smoke outlets. The main object is

LINTEL

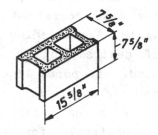

STRETCHER

CORNER

PIER

Figure 22   *Typical masonry units for block-type charcoal kilns*

to stack the wood so that the combustion gases can circulate freely through the pile and most effective use is made of kiln capacity.

Cordwood and slabs are usually hauled to the kiln by truck or tractor, and the pieces are stacked in the kiln by hand. The sticks are commonly piled horizontally, parallel to the sidewalls and on stringers, as shown in Figure 23.

The use of stringers leaves less space for the charge, but better circulation of air and hot gases is thus gained with the result that there is less partly charred material. This material (brands), accumulates usually near the floor of the kiln, where temperatures are lowest. The stringers should be placed so that they cause the minimum obstruction of air intakes and chimney outlets.

Measurement of temperatures is highly important in kiln operation, since the coaling process is controlled by means of temperature and time. The temperature at any given time during the coaling cycle, therefore, gives a direct and reliable measure of the progress of the run. Figure 23 shows details of the thermocouple assembly. Except possibly at the end of the coaling period, smoke colour and volume give little indication of the actual pattern of progress (16).

The heat for initial drying of the charge is provided during the ignition period. This heat is supplied by burning wood fuel placed at midpoint or in front of the charge, or by an oil or gas-fired torch at similar locations.

Some ignition fuels commonly used are dry kindling wood, brands, charcoal, and fuel oil. The amount of fuel required depends chiefly upon the moisture content of the wood to be coaled.

One of the most efficient methods for igniting a charge is with a kerosene or gas-fired torch. These torches are comparatively inexpensive and provide a high-temperature heat source capable of igniting seperate parts of the charge in a very short time. The torch flame is directed through one or more air ports until the charge is burning. Normally, this takes about 5 to 10 minutes (15).

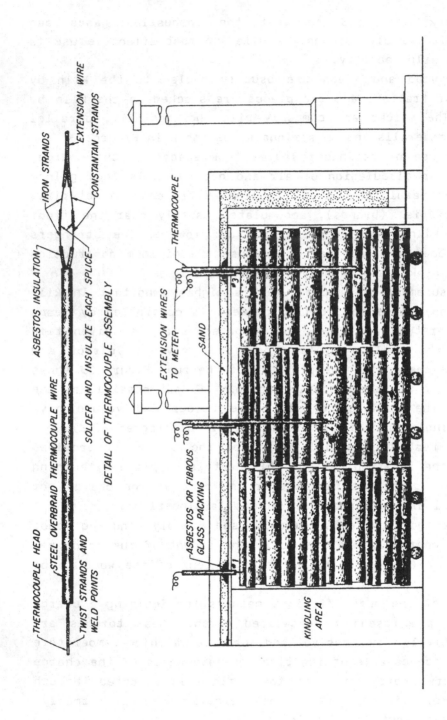

THERMOCOUPLE HEAD

STEEL OVERBRAID THERMOCOUPLE WIRE

ASBESTOS INSULATION

IRON STRANDS

EXTENSION WIRE

CONSTANTAN STRANDS

TWIST STRANDS AND WELD POINTS

SOLDER AND INSULATE EACH SPLICE

DETAIL OF THERMOCOUPLE ASSEMBLY

THERMOCOUPLE

EXTENSION WIRES TO METER

SAND

ASBESTOS OR FIBROUS GLASS PACKING

KINDLING AREA

Figure 23   Detail of the thermocouple assembly on the lengthwise centreline of a cinder-block kiln

(Courtesy of Southeastern Forest Experiment Station Ashville, S.C.)

- 72 -

Satisfactory carbonization depends primarily on maintenance of proper burning conditions in the coaling zone. Sufficient heat must be generated first to dry the wood and then to maintain the temperature necessary for efficient carbonization. At the same time, the burning must be limited so that only sufficient heat is present to produce good charcoal. Kiln temperature is thus the most reliable measure of control.

For the production of good-quality charcoal, kiln temperatures from about 450° to 550° C are required. This charcoal will have a fixed carbon content of about 75 to 82 per cent. Prolonged higher temperatures will reduce the yield of charcoal without necessarily upgrading it for recreational use. If, on the other hand, coaling temperatures remain quite low, the charcoal may be too high in volatiles.

During the coaling cycle, a careful check of kiln temperatures should be made and the air ports adjusted as necessary. The temperatures should be checked at least every 2 to 3 hours for satisfactory control. More frequent checks are advisable when seasoned wood is coaled or during periods of strong or variable winds.

The air supply is regulated by varying the size of the air port openings. Undesirable fast combustion caused by strong winds can be modified or controlled by the use of baffles in front of the air openings. The location of openings for air to be admitted during coaling depends largely on the kiln design and the coaling pattern desired.

In general, coaling time is related to kiln size. When the wood and operating conditions are similar, the time required for coaling in a 35 m^3 kiln will be approximately twice that in one of 18 m^3 capacity. Means for modifying the rather fixed rate of temperature rise to coaling conditions in kiln charges are limited. Attempts to speed up the rate -- for example, by allowing more air to enter -- will raise kiln temperatures excessively and impair charcoal yields and properties (15).

It is possible, however, to adjust coaling time by changing the size of the coaling zone and the manner in which it is directed through the charge. In rectangular kilns it has been possible to reduce the coaling time greatly by centre firing. In the application of this method, the coaling zone widens in opposite directions simultaneously, as compared to the one-direction movement obtained with end firing.

Cooling cycle. When coaling has been completed, all air ports are sealed for the start of the cooling cycle. After the ports are sealed, the chimneys should remain open until smoking has practically stopped. This permits the escape of any smoke that may be formed during cooling and prevents the development of gas pressure in the kiln. Chimneys can usually be sealed from 1 to 2 hours after the air ports have been closed. They should be sealed immediately after they stop smoking, because fresh air may be drawn in by normal cooling, or a downdraught in an idle chimney may admit enough air to support combustion or possibly cause an explosion.

When temperatures have been reduced to 65° C or less, it is generally safe practice to open the kiln. Before the kiln is opened for discharge, however, the charcoal should be checked for localized hot areas that were not evident during the overall kiln temperatures measurements. This check should be made by opening several air ports and one or more chimneys. If the temperature does not rise within 2 hours, it is considered safe to open the kiln completely. If the temperature rises, however, the kiln should be resealed and carefully checked for sources of air leakage.

2.3.3 The Schwartz and Ottelinska Furnaces

Both kilns originated in Europe and were very popular in Sweden. They work on the same principle by using flue gases, for heating. They are constructed with bricks (17).

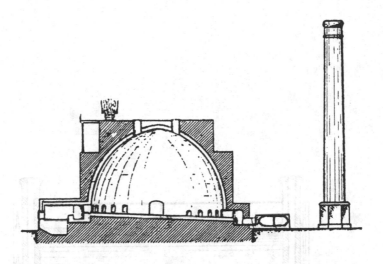

Figure 24   The Schwartz charcoal furnace

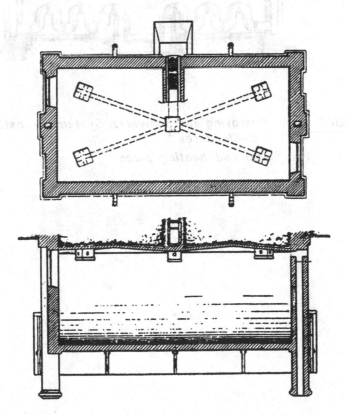

Figure 25   The Ottelinska furnace

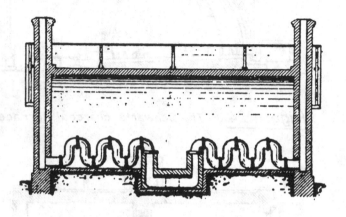

Figure 26   _Improving the Schwartz System by installing_
              „ califorifères "
              (curved heating pipes)

The firing of the kiln is usually accomplished by two burning chambers arranged facing each other (see Figures 24 and 25). The kiln can be fired with firewood or by burning the incondensable gases from the by-product recovery.

The draught for the hot gases is generated by one or more chimneys. The natural draught of the chimneys is supported by bellows.

The off-gas is drawn from the bottom of the kiln through several pipes and is then passed through coolers.

The Ottelinska kiln represents a significant improvement of the technology in so far as the off-gases from the carbonization cycle are drawn from four points located more towards the centre of the kiln.

Usually four or eight kilns form a battery, which is linked together by the off-gas ducts with a central cooling and scrubber system and with a central stack duct.

Further advances have been achieved in the USA by installing so-called "calorifères", which are gas pipes leading the heating gas through the kiln during carbonization (see Figure 26).

During the cooling phase, the cooled down gases from the central cooling and chilling systems are drawn through the same pipe to shorten the cooling time.

2.3.4 The Brazilian Beehive Brick Kiln

These kilns which are operated widely and successfully in Brazil, and especially in the state of Minas Gerais, are internally heated, fixed, batch type. The large iron and steel companies operate several thousand of them.

They are circular, with a domed roof, and are built of ordinary fire bricks. The circular wall is totally in contact with the outside air. This type of kiln is known as the "beehive brick kiln" (see Figure 27).

A variation of the beehive brick kiln is the circular, 4 m diameter kiln that is built into a slope or hill which forms the side and rear walls of the circular kiln. This type is known as the "slope-type kiln" (see Figure 28). It uses far fewer bricks. Many thousand of these kilns are in operation in Minas Gerais and elsewhere in Brazil. They are very popular among the small, independent charcoal producers. Their operation is somewhat easier than that of the beehive brick kilns because there is only one air port to control, as compared with 18 for the regular beehive kilns. Chemical and physical composition as well as the yields of charcoal are comparable to those obtained in beehive brick kilns.

The general data for the two kilns are (18):

	Regular type	Slope type
Kiln diameter volume	5.00 m	4.0 m
Nominal kiln volume	48.94 m^3	24.8 m^3
Effective kiln volume	45.31 m^3	21.6 m^3
No. of air inlet ports	18	1
No. of smoke stacks	6	3
No. of outlet ports	6	4
No. of emergency outlet ports	50	4
No. of bricks	8.500	2.000

Their most advanced modification is presented in Figure 29. The improvement has to be seen in the attached external heating chamber and the reduction of the number of smoke stacks.

For the heating of the kiln usually branches, brushwood and other residue material is burnt, which is not suitable for kilning and would be wasted otherwise.

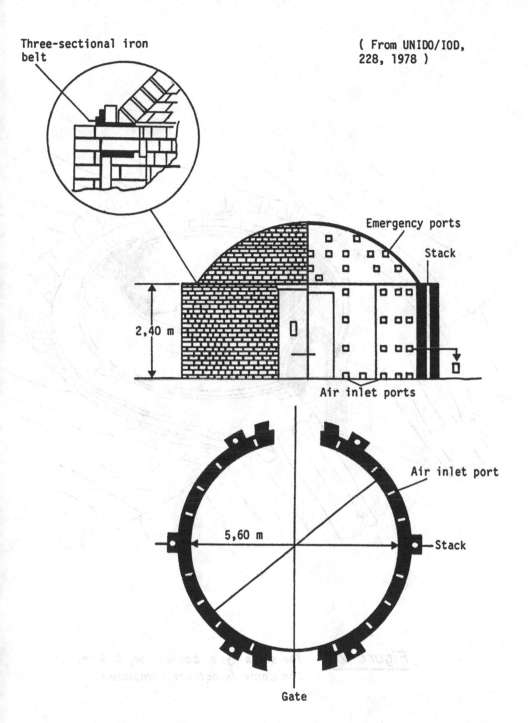

Three-sectional iron belt

(From UNIDO/IOD, 228, 1978)

Emergency ports

Stack

2,40 m

Air inlet ports

Air inlet port

5,60 m

Stack

Gate

Figure 27. The Brazilian beehive brick kiln.

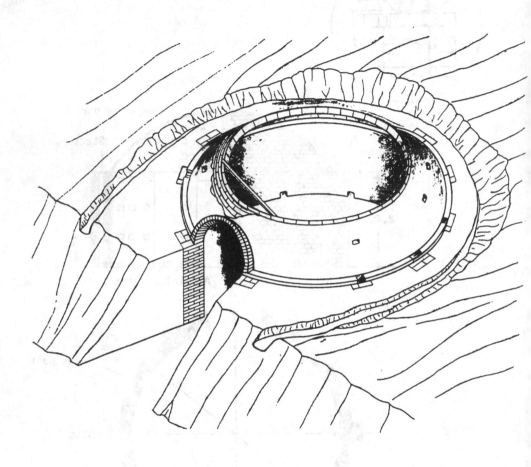

Figure 28   *The slope-type beehive brick kiln.*
The dome is not yet completed.

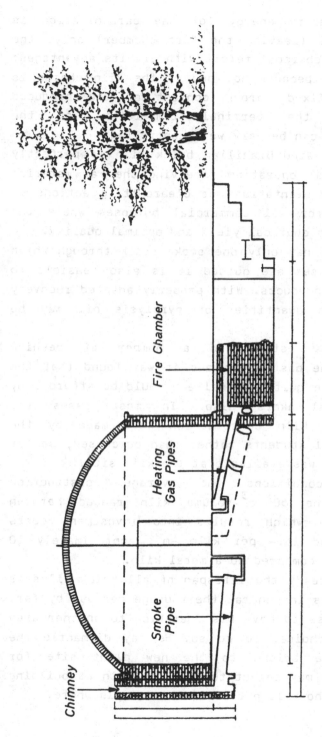

Figure 29   Beehive fire brick kiln with external heating

(Courtesy of Carbon International, Ltd.
Neu-Isenburg, FRG)

Fire Chamber

Heating
Gas Pipes

Smoke
Pipe

Chimney

Since the necessary energy for the carbonization is transferred by gases (leaving the fire chamber) only, the kiln behaves like a charcoal retort with all its advantages: high charcoal yield, because no wood of the kiln charge is burnt away; high fixed carbon content of the produced charcoal, because the terminal temperature of the carbonization cycle can be very well controlled.

The externally heated Brazilian brick kiln is perfectly suited for a charcoal operation receiving the raw material supply from fuelwood plantations or clearcutting actions in forests, that is, for all commercial purposes where the requirements are high charcoal yield and optimal quality.

Because the kiln has only one smoke stack through which the carbonization gases are guided it is also feasible to collect charcoal by-products. With properly adapted recovery equipment attractive quantities of pyrolysis oil may be gathered.

The author has established a number of beehive operations during the past years and it was found that the investment costs were quite nominal and could be afforded by smallholder charcoal-makers also. In most cases the firebricks for the kiln construction were made by the operators or charcoal students rather than purchased, so far adequate clay supply was available at the kiln site.

Under these conditions the average construction expenditures for the 50 m^3-volume kiln ranged between USD 650.- and 800.- which results into investment costs between USD 13.- and 16.- per kiln -m^3, approximately 10 to 15 times lower as compared to a metal kiln.

As mentioned already the lifespan of all brick kilns is very high and exceeds in general their usage periods by far. Should the need arise to move his operation to another area the owner has two choices to do so. He may dismantle the kiln and carry the bricks to the new plant site for reconstruction or he may select to sell the kiln as building material to buyers who will put it in service elsewhere.

Under these view points the investment of a charcoal operation with fire brick kilns becomes even more attractive to the potential charcoal maker.

The construction material for a Brazilian beehive kiln consists of fire bricks, mortar made from clay, a three- or four-sectional iron band for the tightening of the dome and two steel angle lintels for each door.

The building of a Brazilian beehive charcoal brick kiln needs some experience. Therefore, it is advisable for the untaught charcoal-maker and especially for the newcomer in the trade to acquire the necessary skill by training. Preferably, the instruction should be arranged at the envisaged plant site and during the course one or more kilns can be erected and put in use.

Also the operation of the kiln must be learned to obtain an optimal recompense from the invested capital.

The charging of the kiln commences with the forming of a grate on the bottom to allow free gas circulation. The stringers are arranged in such a way that the air can flow freely to the kiln centre.

The fuelwood logs are placed vertically, the thinner pieces against the wall and thicker logs towards the centre. On top of the vertical pile the fuelwood will be placed horizontally until it reaches the ceiling of the dome. All logs must be packed together as close as possible to utilise the kiln capacity to its maximum and to make direct heat transfer easier (Photo 5).

Some kindling is positioned close to the ignition opening which is usually the upper part of the charging gate or the central opening in the dome.

After the ignition white smoke will be issued which turns to a dark colour later. This is considered to be a safe sign, that the kiln has "caught fire" and the ignition opening will be plugged.

The carbonization process proceeds from the top of the kiln to the bottom or from the charging gate to the kiln centre, depending on where it had been lighted.

The operator observes during the entire cycle the smoke issued from the stacks. Carbonization proceeds as long as the colour of the smoke is white or clear (Photo 6).

The draft of the air entering the air inlet ports is regulated by varying the position of brick stones loosely inclined against the porthole entries.

Occasionally a crack may occur in the kiln shell indicated by exiting smoke. This presents no major problem and can be easily corrected by brushing over the leak with a clay slurry which should be always at hand.

Also during the cooling phase the kiln shell will be brushed over with the slurry several times. The number of brushings varies between two and four.

After the kiln has cooled down to 60° – 70° C the kiln is opened rapidly. The skilled operator will smell from the issuing gases whether there is fire inside the kiln. In that case he will extinguish it with a spray of water. Therefore always approximately 200 litres in a drum must be kept ready for use.

It is good during unloading to separate all uncharred pieces of wood. The discharging of the charcoal is done manually and special rakes with wide-spaced prongs are in use.

The discharged charcoal is heaped near the kiln and allowed to cure through aeration for several days.

Fresh charcoal absorbs oxygen eagerly which is accompanied by a rise of the temperature and may cause spontaneous ignition. Therefore, the fresh charcoal is carefully watched by the kiln operator during the curing days. In case of ignition he sprays water over to suppress fire.

For the duration of the cycle of a Brazilian charcoal fire brick kiln no uniform schedule exists, because this

Photo 5
Charging the beehive
brick kiln.
(Photo W. Emrich)

Photo 6   A Brazilian beehive brick kiln in full
             operation.   (Photo W. Emrich)

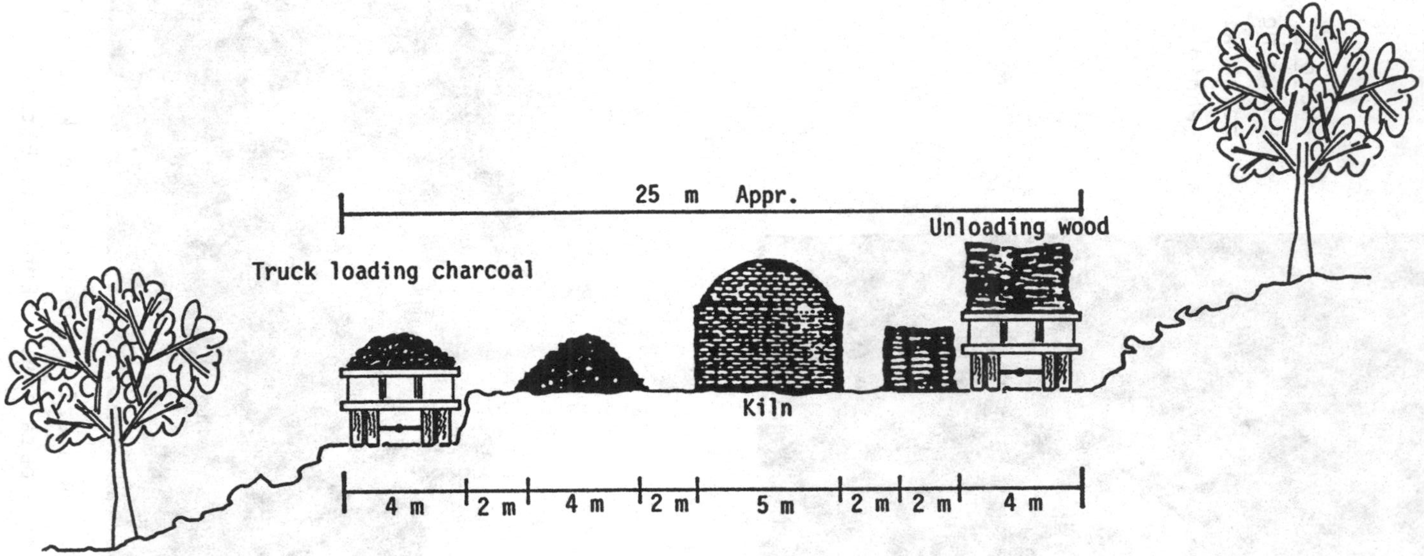

Figure 30. A charcoal production centre.

(From UNIDO/IOD, 228, 1978)

Unloading wood

25 m Appr.

Kiln

Truck loading charcoal

4 m | 2 m | 2 m | 2 m | 5 m | 2 m | 2 m | 4 m | 2 m | 4 m

will be influenced by several factors, including moisture content and diameter of the fuelwood.

The following data have been derived from different commercial operations and may serve the planner and the prospective charcoal maker as a guideline:

Fuelwood loading and charcoal discharging	8 hours
Carbonization	80 hours
Cooling	70 hours
Total cycle	158 hours

The average yield ratio firewood : charcoal is 2.1 : 1 for medium dry wood. For well seasoned fuelwood 2.1 : 1.3 according to many tests the author has performed. The Brazilian charcoal-makers normally apply the ratio 2.1 : 1.

In a slope type kiln in general somewhat lower charcoal yields will be obtained. The firewood : charcoal ratio 2.2 : 1 would be realistic.

With the externally heated Brazilian type kiln a ratio of 2.1 : 1.6 can be achieved, according to own tests.

Commercial operations:

For the production of large quantities a number of Brazilian beehive brick kilns are grouped together in batteries. This allows to simplify the logistics for the raw material transportation and the charcoal pick-up and last but not least it is also labour saving.

The battery consists of seven kilns and several batteries are enclosed in the charcoal production centre.

Practically there are no upper limits for the number of batteries in a production centre except environmental considerations.

During half of the time of the carbonization any kiln without by-product recovery will emit a considerable amount of smoke which can cause embarrassment to residential areas in the vicinity or to the employees working in the centre.

Each battery is attended by two men only, one charcoal operator and a helper.

The centre provides all necessities for efficient charcoal making, e.g. water supply, maintenance shop, stockyard for fuelwood, bagging and loading facilities for the charcoal, etc.

One charcoal battery requires space of the following dimensions: length 70 m and width 30 m.

An important point for the selection of the site is the elevation of the field. An ideal set-up is presented in the Figure 30.

In general the production rates for one kiln battery range between 5.500 and 6.000 m^3 of charcoal per year (19).

2.3.5 The Argentine Kilns

Due to the hemispherical shape the Argentine kiln is also generally referred to as the "Half Orange Kiln". Like any other brick kiln they can be built in large, medium and small sizes. Photo 7 shows a large half orange kiln of approximately 80 m^3 and in Photo 8 a small kiln of only 7 m^3 fuelwood capacity.

A variant of the Argentine kiln is shown in Figure 31 and it is called half orange kiln with "straight jacket", because the kiln cupola rests on the above ground raised foundation (straight jacket).

This kiln type has gained greatest popularity in the charcoal world of South America, especially kilns with a volume of approximately 15 m^3 are frequently employed.

All Argentine kilns are built completely with bricks and in contrast to the Brazilian kiln no iron parts are necessary for construction.

The larger Argentine kilns are designed with two doors whereas the medium size and small kilns have one door only.

The size of the bricks should be 0.24 m x 0.12 m x 0.06 m. But deviations are the rule.

For a large-scale kiln more than 15.000 bricks will be necessary and the medium sized half orange kiln with straight jacket requires not more than 2.500 stones. The mortar is made from clay and water only. To facilitate the dismantling of the kiln later, approximately 10% of charcoal dust are added, which makes the separation of the bricks easier.

Everything which has been said in the previous section concerning the lifespan and durability of a Brazilian kiln can be applied to the Argentine kiln as well.

However, the construction of Argentine kilns requires more skill and the bricks of the cupola must be carefully arranged and oriented.

This skill and the experience can usually be acquired by the charcoal-maker within eight weeks which would also include the instruction time for the operation of the kilns.

Photo 9 shows charcoal students who have almost completed a half orange kiln with straight jacket within one week. The normal construction time for one kiln by two skilled charcoal operators is six days.

The photo also shows the typical technique of arranging the bricks of the cupola by resetting them. To achieve a high kiln volume the distances between each brick layer and the centre point of the kiln are extended starting from the cupola base thus stretching the cupola ark into a more oval shaped form.

The shifting distances depend on the size of the firebricks and can be precalculated. They differ between 0.50 and 6.50 cm from one brick layer to the other.

The following describes the operational procedures for a half orange kiln with straight jacket which the author considers as the best method to obtain good results.

Similar methods of practice may be used to run other Argentine kilns but in each instance they have to be adapted and altered accordingly.

PHOTO 7 The Argentine half-orange kiln. The operator is closing the gate after charging the kiln.
(Photo W. Emrich)

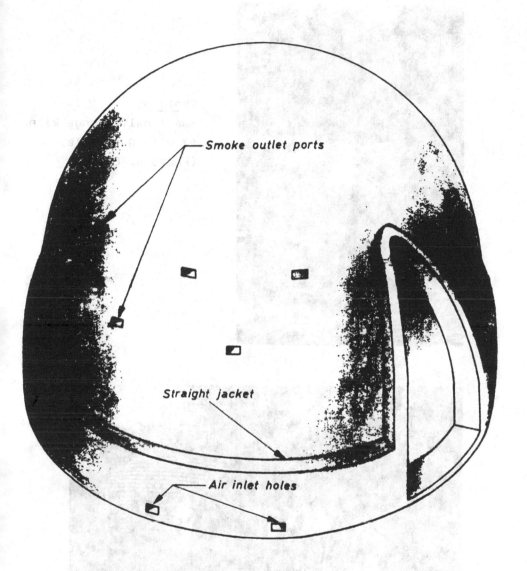

Smoke outlet ports

Straight jacket

Air inlet holes

Figure 31   Half-orange kiln with straight jacket.

Photo 8
Small half-orange kiln
(7 m^3). Guatemala.
(Photo W. Emrich)

Photo 9   Charcoal trainees in Kenya constructing a
half-orange kiln with straight jacket.
(Photo W. Emrich)

Photo 10   Charcoal trainees igniting the kiln with a
            shovelful of glowing charcoal. Kenya.
            (Photo W. Emrich)

Photo 11
Charcoal trainees
brushes over leaks.
Kenya.
(Photo W. Emrich)

Operation of the half-orange kiln

Fuelwood

The fuelwood is cut into pieces approx. 1.00 m - 1.30 m in length (minimum diameter of 5.0 cm, maximum diameter of 50.0 cm).

The fuelwood which is transported to the kiln site should be stored as close as possible to each kiln. A minimum of five to six weeks air-drying time is recommended.

Thicker logs -- e.g. with a diameter in excess of 20 cm -- should be split once or twice to facilitate the reduction of moisture content and to shorten the carbonization time.

Charging the kiln

First of all, stringers are placed on the kiln floor. This is to prevent direct contact between the fuelwood and the ground and to provide sufficient space to allow free circulation of the air from the inlet holes through the centre of the kiln. The logs are stacked vertically on top of the stringers and packed as tightly as possible.

All logs whose diameter exceeds 25 cm are positioned in the centre of the kiln so that they will be exposed as long as possible to the higher carbonization temperature.

The kiln charge is completed by the addition of a layer of logs which is stacked horizontally on top of the vertical logs until reaching the ceiling of the cupola.

Special care must be taken to ensure that the air inlet holes at the kiln base remain open.

Some dry wood or rubbish is placed in the upper part of the kiln door to serve as kindling.

When the kiln has been fully charged, the door is closed with bricks and mortar and covered with mud from the outside, except for an opening measuring approx. 20.0 x 20.0 cm in the upper part of the door. This is called the "ignition eye".

Ignition of the charge (see photo 10)

All inlet holes and smoke ports must be open. One or two shovels full of glowing charcoal are thrown in through the ignition eye. Initially, the kiln will give off bluish smoke, which turns white after a short time. This indicates that the initial phase of distillation has begun and that the fuelwood is losing moisture content. At this time the ignition eye will also be filled in with bricks and mortar and firmly sealed.

As a rule, the time between ignition and the closing of the eye -- when the charge has caught fire -- does not exceed twenty minutes.

The carbonization phase

The white smoke will continue to be given off through the upper smoke ports for several hours and then start to turn bluish. As soon as blue smoke is released from a particular smoke port, the operator closes this port with a brick fitted to the opening and seals it with mud mortar.

There is no set rule as to where the smoke colour change from white to blue will occur first -- e.g. one cannot predict which smoke port will be the first to emit bluish smoke -- and this may also depend very much on the prevailing wind direction at each individual kiln location. Furthermore, the change in colour will not occur simultaneously in all smoke ports; rather, one port after another will begin to discharge blue smoke. After the upper smoke ports of the kiln have been closed and properly sealed, the white smoke will be released through the lower row of smoke ports only. In closing the lower ports, the operator follows the same procedure as in the case of the upper ports, carefully monitoring the colour of the smoke.

As soon as the smoke from a particular smoke port has clearly turned bluish, the operator uses a stick to probe inside towards the centre of the kiln to ascertain whether or not there is an obstruction (uncarbonized wood). If there is no impediment, this smoke port may be closed. If uncarbonized or partly carbonized wood blocks up the path of the probe stick to the centre, the hole may be partially closed using a specially fitted brick. But under no circumstances should it be sealed completely.

This procedure serves to delay the combustion of the charcoal in the vicinity of the smoke port and enhances carbonization.

If the hole is re-checked within one hour and the second probe indicates that a significant amount of uncarbonized wood is still present, it is advisable to slow down the air influx by partially closing the two nearest air inlet holes at the base of the kiln using brick "stoppers".

After all of the lower smoke ports have been closed and properly sealed, smoke will also begin to exit through some of the air inlet holes.

This is perfectly normal and all monitoring and operational producers must be followed until the last air inlet hole has been closed and sealed, at which point the cooling phase begins.

If the kiln has been properly charged and operated, the carbonization phase should be completed some ten to twelve hours after ignition.

The cooling of the kiln

It is important that the kiln shall be airtight and exhibit no leaks or cracks through which air could enter. If air is present, the charcoal charge will start burning and cooling will be delayed significantly.

Therefore, after the kiln has been closed and sealed, at least three coats of mud slurry are applied to the exterior surface (see Photo 11). This will also help to reduce the cooling time. When the kiln has cooled down sufficiently, the door may be opened and the fire extinguished with water; one drum of about 200 litres is sufficient for one kiln. After the fire has been completely extinguished, discharging may begin.

The kiln is discharged by two or three men using rakes. These special tools have 12 - 14 prongs spaced 2.0 cm apart. Their employment allows the bulk of the fines (less than 20 mm) to fall through and remain in the kiln. They may be removed later after the interior of the kiln has cooled down further.

The charcoal is then hauled to the nearby storage area for curing. The most common method is simply to place the charcoal on a piece of canvas, which is then carried to the storage area by 3-4 men.

Charcoal should not be handled or transported in large quantities until sufficient "curing time" has elapsed following discharging. A curing time of eight to ten days is usually considered sufficient to quell the self-igniting tendency of freshly produced charcoal.

During curing, neither the height nor the diameter of the charcoal heaps should exceed 1.50 m otherwise exposure to the air will be insufficient.

Maintenance of the kiln

The structure of the Half-Orange kiln can be damaged, for instance by the impact of logs, and this should be avoided. Bricks which have fallen out of the walls or have become loose should be put back in place and rammed tight.

Periodically, the excess clay which has accumulated on the exterior of the kiln as a residue of the successive brushings with clay slurry should be removed with a rasp. This accelerates the charcoal cooling process.

The kiln floor should always be kept level. If necessary, some wet clayish soil should be put in and stamped down. The water drainage ditches around the kilns must always be kept unobstructed and clear of all rubbish.

The economics

Very little has been made public about the efficiency of Argentine kilns.

The yearly average obtained by the largest charcoal producer in Argentina, Salta Forestal S.A., was 3.75 tons of fuelwood per ton of charcoal (20) in 1978, resulting in an average yield of 26.67% (weighted bases). This figure, however, applies to large Argentine kilns with a bottom diameter of 6.00 m.

Extensive studies carried out on a commercial scale for East African countries proved an efficiency for the half orange kiln with straight jacket of 28.2 % (21). The carbonization tests were carried out with acacia species as fuelwood.

Based on the same test series the following operational data were extracted:

```
Annual charcoal output      3.500 tons
Number of kilns                28
Investment cost*      USD  15,600.-  = USD 558.-/kiln
Kiln operating cost   USD  14,300.-  = USD 4.14/t charcoal
```

A typical cycle has been established as follows:
- charging of fuelwood   5   hours
- ignition of kiln   (30   minutes)
- carbonization   18   hours
  (measured between ignition and
  closing of 80% of all kiln apertures)
- cooling phase   22   hours
- unloading of charcoal   3   hours
                                         48.5 hours

---

*   If bricks are made by the charcoal operation the costs would be cut in half

2.4 Kiln Designs for Waste Conversion

During the last decades many attempts have been made to carbonize forestal and agricultural waste with simple and inexpensive kilns.

Since most of the waste matter is found or being discharged in small fragments and particles the kiln technique described in the previous sections is not suitable for waste conversion without modification.

If a large kiln is charged with fragmentary matter, for instance with saw dust, the denseness and compactness of the kiln load will not allow sufficient penetration of gases, unless the cargo is moved or rotated continuously.

With the details of waste conversion and their proper solutions will be dealt within the next chapter.

Another view point frequently overlooked is the fact that the carbonization of forest and agricultural waste will produce mainly charcoal fines which cannot be used for household fuel, as an example, and the charcoal-maker must consider an agglomeration or briquetting plant also.

Naturally, this will imply higher investment cost especially if the targets are export markets with their elevated charcoal quality standards. Their specifications can be met only with specialised and expensive machinery and equipment.

Although the requirements of local markets in developing countries are more lenient and less expensive outfits can do a satisfactory job, the operational costs for charcoal briquettes or formed charcoal are perceptibly higher as compared to the production of lump charcoal.

Consequently, many charcoal producers try to set off their extra expenses with the gains from sale or by utilisation of by-products.

It is not possible to describe within the frame of this handbook all modified systems which the igenuity of numerous charcoal-makers has brought forth during the past ten years alone. Instead, one scheme will be presented in detail,

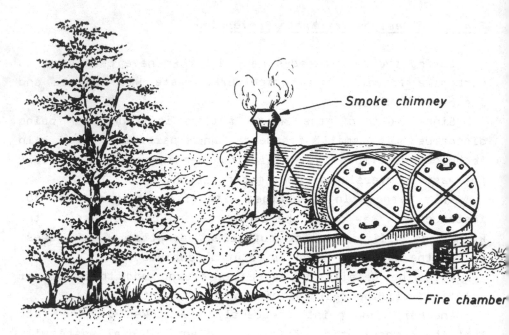

Smoke chimney

Fire chamber

Figure 32 *Carbo-Gas retort (twin unit). For better insulation the retort may be covered with sand or clay.*

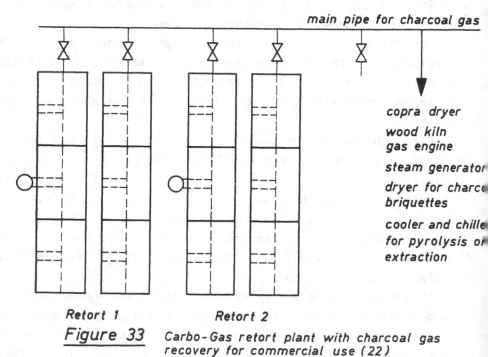

main pipe for charcoal gas

copra dryer

wood kiln
gas engine

steam generator

dryer for charc
briquettes

cooler and chille
for pyrolysis or
extraction

Retort 1   Retort 2

Figure 33 *Carbo-Gas retort plant with charcoal gas recovery for commercial use (22)*

which the author has subjected to several trials and has proved commercially successful.

2.4.1   The Carbo - Gas Retort

The retort is one of the most efficient means to produce high quality charcoal and to collect carbonization by-products.

The retort presented in Figure 32 can be constructed from oil drums. Twice three oil drums assembled form one carbonization unit above one fire chamber. An extra drum can be added if necessary.

At the rear end of the retort a pipe is installed for collection of the charcoal gas. In the simplest way the pipe leads the gas underneath the retort where it is burnt in the fire chamber.

In other designs the retort gas is guided into a main pipe from where it can be drawn-off to various users. The Figure 33 demonstrates a larger Carbo-Retort operation with possible usages for the charcoal gas.

The retorts can be set up over a trench if the earth is solid enough. In this case bricks for the support are not necessary.

In general the kiln units are raised above ground by a brick construction which also serves to make the fire chamber.

The retort operation must be installed where the by-products can be utilised or where sufficient demand for such by-products can be found in the vicinity.

Raw material supply and preparation

The feed for the Carbo-Gas retort may be found in forests, sawmills, furniture plants, plantations, food processing factories, copra drying stations, etc.

Some of the waste matter is very moist. The water

content can be reduced by kiln drying using the charcoal gas emitted by the process for heating.

Size reduction of the waste matter is usually not necessary. Branches or cotton sticks, as an example, can be hogged or broken in small pieces by simple means.

The prepared raw feed is loaded on curved trays, which fit the cross-section of each retort as much as possible.

These trays are made from sheet metal and are left in the retort until the charcoal is ready.

Operation (22)

Early in the morning the retort will be charged with the prepared feed trays. It is advisable for large operations to have always an extra set of filled trays sitting next to the kilns ready for loading.

After the covers are screwed on to the bolts the fire under the retort (leaves, bark, branches etc.) is made. The flames should touch the retort below.

The carbonization starts when the gas flows through the pipe, usually after one or two hours. The process has terminated when the gas flow ceases.

If the gas evolving from the retort is burnt in the fire chamber the fire must not be stoked repeatedly. The charcoal gases will burn and maintain the temperature until the end.

The retort remains tightly closed overnight for cooling. The next morning the cover of the retort will be unbolted and the tray with the charcoal is pulled out with a hook which fits into a hole in the tray.

In larger operations always half of all retorts are in use and will produce gas. The gas collected by the main pipe, therefore, has an almost uniform consistence and represents an exellent fuel for electric generators also.

The produced char has to be stored for curing and then may be processed.

Under the heading of chapter 8 the details for making charcoal briquettes and extrudates are described.

Economic considerations

As with any charcoal retort the conversion efficiency is very high. In the case of waste conversion, the quality of the produced charcoal will depend very much on the ingredients of the raw material. Rice husks for instance contain significant amounts of incombustible minerals. They will accumulate during the carbonization in the char. This effect may exclude all charcoal made of rice processing residues from many industrial applications.

In contrast, carbonization products made of wood waste or nut shells have excellent properties (23).

The investment cost of the plant will also depend very much on the local situation, that is price of metal scrap, maintenance cost, etc.

Finally the raw material cost cannot be calculated on a general basis. They will differ on the fact whether the waste and residue materials have already a market value or not. Many are used to fuel a factory, others are available in excess, and they have to be disposed of which would cause additional expenses to the owner.

Retort operations can be designed as a small business or they can be planned and organized for large-scale production and fitted into a whole integrated industrial concept.

An important role for the plant economics play the gains achieved by the utilisation of the gaseous or liquid by-products. Approximately 40% of the organic matter put into a retort process can be recovered as gas or vapors.

In some circumstances the sales will pay for the operational costs entirely.

Each factor plays a part in making the venture work and the business go. But they cannot be discussed in detail without sufficient information of the local particularities.

References

(1)   J.I. Spaeth, <u>Praktische Abhandlung über das</u>
        <u>Verkohlen des Holzes in grossen und kleinen</u>
        <u>Meilern</u>, Nürnberg, FRG, 1809

(2)   D. A. Tillmann, <u>Wood as an Energy Resource</u>,
        New York, USA, 1978

(3)   K. Openshaw, <u>Costs and Benefits of Proposed Tree</u>
        <u>Planting Programme for Satisfying Kenya's Wood</u>
        <u>Energy Requirements,</u> Stockholm, Sweden, 1982

(4)   M. Vahram, <u>Quality of Charcoal made in the Pit</u>
        <u>Tumuluts</u>, National Science Research Council,
        University of Guyana, Charcoal Unit Laboratory,
        Report Nr. 4

(5)   <u>FAO Forestry Paper 41, Simple Technologies for</u>
        <u>Charcoal Making</u>, Food and Agriculture Organization
        of the United Nations, Rome, Italy, 1983

(6)   H. Bergstrom, <u>Handbook for Kolare</u>, Stockholm,
        Sweden, 1934

(7)   H. Bergstrom, <u>Om Traekoling. Toll tjaemst foer</u>
        <u>Undervisningen vid skogs - och kolarsbar samt</u>
        <u>tekniska Undervisningeanstalter och foer praktisk</u>
        <u>bruk</u>, 2. Auflage, Stockholm, Sweden, 1918

(8)   A. Klanins, <u>Die Holzteerschwellung</u>, Riga, UDSSR,
        1934

(9)   F. Klein, <u>Über das Verkohlen des Holzes in</u>
        <u>stehenden Meilern</u>, Gotha, G.D.R, 1830.

(10)   Research work carried out by the Forestry Technical
         Department of the Colonial Ministry, Paris, France,
         1941 - 1943

(11)   Publication of Appropriate Technology Development
         Institute, Wokim Sakol Long Drum, (in Pidgin
         Language), Lae, Papua New Guinea

(12)   W. Emrich, Charcoal Making in the Philippines,
         Neu-Isenburg, F.R.G., 1982

(13)   Whitehead W D J, The Construction of a
         Transportable Charcoal Kiln, Rural Technology
         Guide, Tropical Poducts Institute No. 13, 1980

(14)   J.P. Jarvis, The Wood Charcoal Industry in the
         State of Missouri, University of Missouri,
         Columbia, USA, 1960
         Anonymous, Charcoal Production in Kilns, Forest
         Product Journal 7 (110, 339-403), 1957

(15)   Anonymous, Production of Charcoal in a Masonry
         Block Kiln. Structures and Operation, Forest
         Service, US Department of Agriculture, Processed
         Report No. 2084, 1957
         Anonymous, Facts about Concrete Masonry, AIA File
         Nr. 10-c, National Concrete Masonry Ass., Chicago,
         III., USA, 1947

(16)   P. Ralph, An Inexpensive Method for Measuring
         Charcoal Kiln Temperatures. Southeastern Forest
         Experiment Station, Ashville, N.C., USA

(17)   H.M. Bunburry, The Destructive Distillation of
         Wood, London, UK, 1925

(18)   H. Meyers, <u>Charcoal Ironmaking, a Technical and</u>
         <u>Economical Review of Brazilian Experience</u>,
         UNIDO/IOD. 228, 1978

(19)   Fundacao Centro Tecnologico de Minas Gerais
         (CETEC), Belo Horizonte, Brazil

(20)   M.A. Trossero, <u>Analysis Comparativo de Hornos de</u>
         <u>Carbon Vegetal</u>, Congreso ILAFA-Altos Hornos,
         Instituto Latinamericano del Fiero y el Acero, 1978

(21)   W. Emrich, <u>The Feasiblity of Charcoal Making in</u>
         <u>Selected Areas of East Africa</u>, Carbon
         International, Ltd., Neu-Isenburg, F.R.G., 1984

(22)   W. Emrich, <u>Making Charcoal from Forestal and</u>
         <u>Agricultural Waste the Retort Way</u>, Carbon
         International, Ltd., Neu-Isenburg, F.R.G., 1984

(23)   Paper # 63, <u>Coconut Shell Charcoal</u>, Food and
         Agriculture Organisation of the United Nations,
         Rome, Italy

Chapter 3
CONCEPTS AND TECHNOLOGY FOR THE INDUSTRIAL CHARCOAL-MAKER

Industrial charcoal-making has a comparatively short history dating back about 150 years. The principles may be outlined as follows:

- Relatively high investment costs;

- Intensive use of labor-saving equipment and devices;

- Efficient recovery of liquid and/or gaseous by-products for captive and commercial use;

- Wide range of raw material usage, including forestry residues as well as agricultural and municipal waste;

- Such projects necessarily involve prior feasibility studies, qualified plant design and organisation of logistics.

According to these criteria the Missouri kiln (see 2.3.1), the Brazilian beehive brick (2.3.4) and the Argentine kiln types (2.3.5) can also be considered as industrial charcoal technology when employed in large numbers.

3.1 Equipment for Charcoal Plants with By-Product Recovery

By the middle of the nineteenth century, the potential value of the by-products of charcoal plants had become obvious. The emergence of the chemical industry brought about a pressing need for the supply of organic acids, methanol and acetone, compounds which were present in the condensates of charcoal plants. The problem of recovery and refining them could be resolved to some extent.

The charcoal-makers of this time immediately recognized the new trend and its potential, and they started to adapt the existing plants and/or if necessary established new concepts. The equipment changed radically from old-fashioned kilns to modern retorts with auxiliary installations for the recovery of by-products (1).

The change in the charcoal industry precipitated a mushroom growth of patents during the following seventy years. It was only interrupted by the Second World War when the charcoal industry was faced with the task of helping to sustain the war machinery. Some of the new charcoal technology of these times is still in use today, demonstrating its fundamental value.

The charcoal planner needs to know the essential techniques and to understand the elements of industrial charcoaling. The status of the art of modern charcoal technology may be summarized as follows:

- Operation of large-scale retorts with a capacity of 100 m^3 and more for the charcoaling of wood logs.

- Continuously operated large- and small-scale retorts/converters for the carbonization of wood logs and forestal and agricultural waste.

- Improved equipment for the recovery and fractionation of pyrolysis oil.

- Mechanised equipment for the charging and discharging of charcoal retorts and converters.

- Advanced technology for the briquetting and agglomeration of charcoal.

- Integrated carbonization concepts utilizing the energy content of raw materials at an optimum level.

The most revolutionary step forward was made by the introduction of equipment for the charcoaling of biomass in the broadest sense. Until then, roundwood, thick branches of trees from natural forests were the only choice. With the first appearance of continuously run vertical retorts, the residues and waste discharged by sawmills, plantations and food processors also became important for the charcoal-maker.

Biomass converters were first developed for very large plants with a charcoal production capacity exceeding a thousand tons per month. The equipment was borrowed from other industries, e.g. fertilizer and cement plants, and adapted to charcoaling.

Since the installed plant capacity was high, huge accumulations of raw feed, e.g. sawdust, bark, nutshells, were a necessity for economic operation. In many cases, the collecting of waste and residues proved uneconomic and was therefore abandoned.

After the new furnaces had been employed by large charcoal plants for a while, the small charcoal-maker also became interested also, at first in North America. Commercial plants were erected between 1960 and 1965 and soon proved their economic feasibility.

For the classification of charcoal technology, this handbook employs a modern terminology which differentiates between traditional charcoal-making (kiln technology) and the industrial processes. In this context, kiln technology means all equipment producing charcoal as the sole product, whereas retorts or converters are essential elements of the industrial plant, capable of recovering and refining charcoal by-products in commercial grades and quantities.

3.1.1 The Forerunners of Modern Charcoal Equipment

The Reichenbach and Carbo furnaces

The first of these, once in widespread use in Finland

and Sweden, has a brick-built chamber heated by hot gases circulating in an outer casing. The cylindrical chamber, of 30 to 40 steres capacity, is surrounded up to two-thirds of its height by a second wall which forms a circular heating chamber. The brickwork often shows cracks and part of the volatile products burn in the annular heating chamber. Heat output is poor.

The operation lasts about 6 days including loading, distillation and cooling. The distillation products are evacuated at the base by a copper pipe shaped at the end like a truncated cone (1, 2).

The furnace yields the purest kind of tar and a good quality crude pine oil. It produces no pyroligneous acid.

The CARBO kiln derived from the above is, however, a much improved version (see Figure 34). It combines external heating with internal tubular heating. It is a vertical cylinder in sheet metal of 300 to 400 steres capacity. In the middle of the slightly conical base, there is a collecting conduit for by-products. The apparatus has a metal lid with four loading apertures. The charcoal is unloaded from a door in the base. Working is discontinuous (1).

The gases from the external furnace adjoining the chamber circulate through spiral tubes surrounding the chamber; there are dampers to regulate the heat. The lower third of the retort is protected by firebrick so, that it will not be damaged by the hot gases; the gases react directly upon the metal casing of the top two-thirds. A special feature of the CARBO is a large vertical pipe passing through the middle of the chamber and divided lengthwise by a partition which does not reach the base of the pipe. The base of the pipe is devised so that the non-condensable gases and the air required for their combustion can enter. This air is supplied by the excess external heating gases which, issuing from the spiral conduit, are carried into the left-hand section of the central pipe. Hence the combustion of the non-condensable

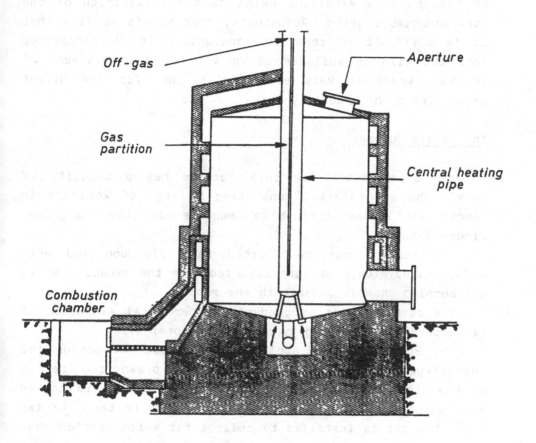

Off-gas

Aperture

Gas
partition

Central heating
pipe

Combustion
chamber

Figure 34   *The Carbo furnace.*

gases takes place starting from the base of the centre pipe; after combustion, the gases escape into the air (or into a regenerator) through the right-hand section. The heat output of this kiln is excellent owing to the utilization of the non-condensable gases. Adjustment, however, is so fine that it is difficult to regulate constantly; it is considered that the kiln is suitable above all for the treatment of resinous woods of very even quality and for the direct extraction of high-quality tar.

The Bosnic furnace

The steel retort of this furnace has a capacity of $50 m^3$. The raw material consisting of logs of wood 1 m in length is loaded through a manhole in the top (see Figure 35).

The heating can be started with firewood and when off-gases develop, they are directed from the exhaust fan to the burning chamber underneath the retort.

The carbonization is conducted under total exclusion of air according to heating type B (see Chapter 1).

The Bosnic Furnace applies the downdraught scheme for the off-gases, meaning that the gases are drawn from the top of the retort through the bottom. The hot gases are cooled with water and the condensed pyrolysis oil is taken to two vats. One vat is installed to collect tar which settles down after a while (2).

A smaller but very effective design of the Bosnic Furnace is shown in Figure 36. The principle of the technology applied is strictly the same as in the larger operation.

The retort capacity is 5 to 7 m^3 to facilitate exchanging, which is conducted by manual hoisting. The raw material is again roundwood or split logs which must be packed very tightly in the retort. When the carbonization cycle is terminated, the hot retort is hoisted off and set

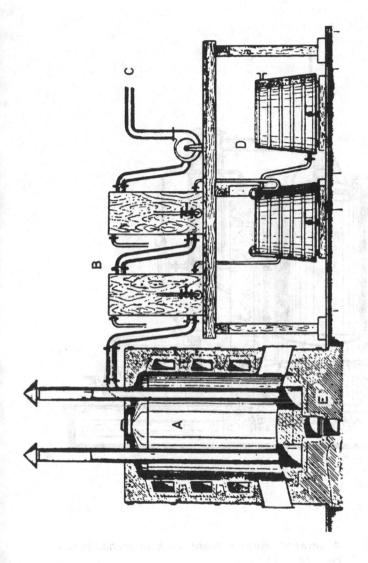

A = Retort
B = Off-gas cooler
C = Gas fan
D = Pyrolysis oil tank
E = Combustion chamber

Figure 35   *The Bosnic charcoal plant*

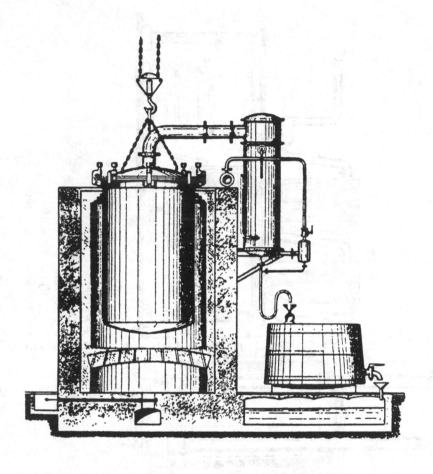

Figure 36   A smaller Bosnic plant with interchangeable retorts.

aside for cooling the charcoal inside. The burning chamber is now ready to take in the next retort with a fresh charge. The changing is done quickly to prevent the burning chamber from cooling-off, thus saving fuel.

These plants demonstrate two features of industrial carbonization which had been neglected by the traditional charcoal-maker until then

- Increase in the energy efficiency of the charcoal plant by better utilization of the raw material.

- Diversification by recovery of by-products for commercial use and recycling of residual gases for heating the plant.

The structure of these charcoal plants is relatively simple and no exotic construction materials are required. Retort steel, firebricks, wooden vats are the essential constituent elements. Only in sensitive areas within the off-gas cooling system must copper or stainless steel be employed.

The Badger-Stafford process

This process was formerly used by the Ford Motor Company for dealing with certain sizes of factory wood waste (birch, oak, maple, ash). The waste was cut into pieces 20 x 5 x 2 cm and dried in a revolving drum 3 m in diameter and 30 m long. Drying, effected by means of a counterflow of furnace gas from the ovens of the steam-generation plant, lasted 3 hours. Output was continuous. Six drums were capable of dealing with 400 tons of wood per day (3).

The wood unloaded from the drums at 150° C contained only 0.5 per cent moisture. The wood was introduced mechanically into a chamber with 3 shoots which distributed the load into 3 Badger-Stafford retort works continuously, for 2 weeks, after which the encrusted tar had to be burnt out; hence, of 3 kilns, 2 were working while one was laid up for cleaning.

The charcoal which settled down continually in the retort cooled to 255° C at the base of the retort. It was then de-kilned into a revolving drum cooled by outside water circulation. The drum was 1.8 m in diameter and 9 m long. Finally, the charcoal was brought into contact with the air in a revolving cylinder where it was stabilized by recovering oxygen. The complete operation of cooling and curing took only 5 hours.

The average yields in percentage of the weight of air-dry wood were:

Charcoal:	20.0 %
Wood spirit:	1.7 %
Acetic acid:	4.5 %
Tar:	10.0 %

The powdered charcoal was made into briquettes. Precautions would have to be taken against the risk of explosion, especially the thorough removal of dust from all gases.

Despite its remarkable features the Badger-Stafford process proved uneconomic and the plant was shut down soon.

3.1.2 Retort Technology

Although there is no precise demarcation between charcoal retorts and converters, the latter term is applied to equipment capable of carbonizing biomass fragments and small particle-sized waste. Consequently, the term "retort technology" refers to carbonization of pilewood or wood logs reduced in size to a minimum length of 30 cm and not exceeding 18 cm in diameter. For simplicity, the traditional route will be followed and therefore the retort and biomass technologies will be discussed under different headings.

3.1.2.1 The Wagon Retort

The wagon retort process was once of great commercial importance in European countries and in the USA. More than 70 % of all the charcoal produced was made this way. The process has lost its significance, mainly because of its high manpower requirement compared with other methods. However, one or two plants have survived the waves of rationalization even in Europe with its high labour costs.

Process description

The principle of a wagon retort plant is illustrated in Figure 37. The most commonly utilized raw materials are roundwood, split roundwood and sawmill waste with an average length between 1.0 and 1.2 m. Shorter pieces may be charged also, but in limited quantity.

The prepared wood logs are loaded into small lorries which have a grate on the top. The shape of the grate is designed to fit the cylindrical retort tightly, and to make optimal use of the retort space.

A rail system connects the wood storage area with the plant site. The loaded lorries are either pushed manually or jerked mechanically into the retorts. Normally, two lorry charges will fill the retort.

Large plants have a shunting system for the lorries, with shunting-engines and sidings extending for several kilometres.

The cylindrical retorts have a standard length of 8 to 9 m, and their diameter ranges up to 2.50 m. They are made of steel and are inserted in a brick structure, with ducts for the heating gases (3).

All the known wagon retort operations employ external heating systems. The off-gases of the retort are drawn-off by a fan and after devaporizing (pyrolysis oil recovery) are directed into a burning chamber underneath the retorts.

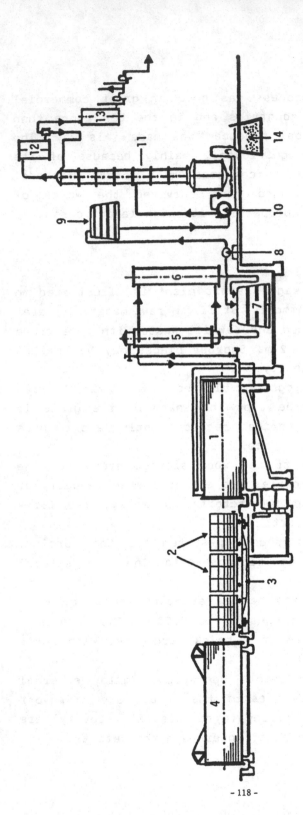

1 = Retort
2 = Lorries
3 = Siding
4 = Charcoal cooling cylinder
5 = Water cooler
6 = Scrubber for residual gas

7 = Pyrolysis oil vat
8 = Oil pump
9 = Intermediate tank
10 = Preheater for distillation column
11-13 = Fractionation of pyrolysis oil
14 = Settling vat for tar

Figure 37. The wagon retort plant.

The carbonization time totals one working day. After the coaling phase has terminated, the lorries are pulled from the retort and pushed with the glowing charcoal into the cooling cylinder. To facilitate the lorry transfer and to prevent vigorous burning of the charcoal when it comes into contact with the air, these coolers are positioned in front of each retort cylinder. The cooling of the charcoal to the point where it is no longer self-igniting can be shortened by spraying water on the cooler cylinder.

The off-gases are passed through a cooler and scrubber to strip the pyrolysis oil from them. In our illustration, a fractionating column is attached to the plant for the recovery of raw acetic acid from the oil. This addition is not necessary if the pyrolysis oil is to serve only as a fuel.

Operational data

- Usable retort space: length 7.50 m, diameter 2.50 m

- Capacity: Approx. 35 m^3.

- Possible throughput: 9 tons/m^3/month, dry basis.

- Raw Material: Wood, log size 1.0 to 1.20 m, max. diameter 8-12 cm. Moisture content max. 25 % (wet basis); wood species, no requirements.

- Average Yield: charcoal 33-38 %, pyrolysis oil 20-25 %

- Energy demand: 0.6 to 0.75 million kcal/ton raw material, approx. 80 % covered by devaporized off-gas. 18-20 KWh/ton raw feed. If no pyrolysis oil is recovered, the energy supply from the off-gases alone will be sufficient to cover the demand of the retorts and the heat for the briquetting as well.

-119-

- Retort structure: Riveted metal sheets, gates and framework of cast iron, sealing of the gates for the retorts and charcoal cooler by asbestos.

- Economic considerations: A battery of six to eight retorts is considered to be an economic plant size. The logistics for the wood supply, storage and preparation of the wood lay a heavy burden on the operational cost. Maintenance costs of lorries are high. To set-off expenses, the recovery of pyrolysis oil is desirable. The operation is simple, requires little personnel training, and the installed plant equipment is robust, withstanding misuse.

- Disadvantages: Mechanical drying of raw material is not possible, air drying by stacking is the only way; consequently, a large storage area is required and capital investment is increased. Cracking of tar causes crusts on retort walls which slow down heat exchange; the operation must be interrupted frequently in order to clean the retorts.

3.1.2.2 The Reichert Retort Process

The centre piece of this process is the large-scale retort with a raw material capacity of 100 m^3. The batchwise process requires wood logs or sawmill waste which must be reduced in length to a third of a metre.

Another significant feature of the process is the heating system which belongs to type C (Chapter 1.3), utilizing recirculated off-gas.

The principles of the process are shown in Figure 38. The wood is charged at the top of the retort through a manhole, up to capacity. After closing the valve, the main heating pipe is opened to let hot gases into the interior. Since the retort is designed for downdraught, the carbonization starts in the upper layers of the charge where the hot gases enter. During the cycle, the charcoaling zone moves slowly down to the bottom.

On its way out, the off-gas passes through the uncarbonized feed, taking off the moisture.

After leaving the retort, the off-gases are first stripped from tar and then chilled to obtain the pyrolysis oil. The residual gas is scrubbed and introduced into the central burning chamber. Part of the gases are burnt to raise the temperature of the heating gas leaving the retorts to the original level of 450° C.

After carbonization has stopped, the charcoal is released at the bottom of the retort. To prevent losses by self-ignition of the hot char, it is dropped into airtight containers. The cooling is usually done by passing cold inert gas through the charcoal.

Large-space retorts with recirculated gas heating have been in commercial use for more than forty years. In countries with high labour costs, they will work economically only if they are provided with labour-saving mechanical installations for the conveying, loading, discharging and preparation of the raw material and charcoal. High investment costs have prevented their widespread use until now.

Operational data

- Retort size: height 8.50 m, diameter 5.0 m, vertical.

- Capacity: approx. 100 m^3.

- Possible throughput: 8 tons/m^3/month, dry wood.

- Raw material: Wood logs, max. length 35 cm, max. diameter 15 cm, moisture not exceeding 20 %.

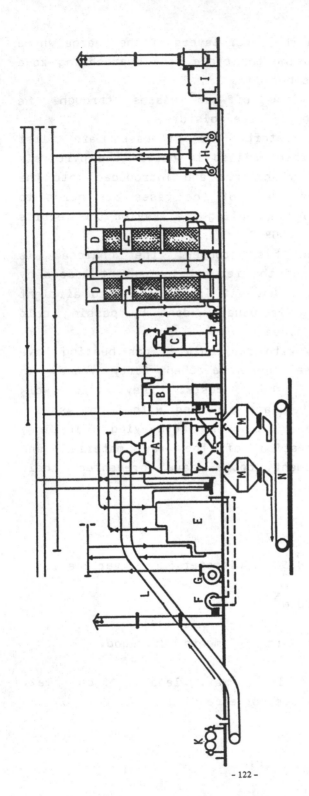

A = Large Space retort (100 m³)
B = Tar stripper
C = Water cooler
D = Scrubber for residual gas
E = Combustion chamber and heat exchanger
F = Off-gas fan
G = Fan for combustion air

H = Fan for forced-gas circulation
I = Dust collector
K = Circular saw for wood preparation
L = Conveyor for retort charging
M = Charcoal cooler
N = Conveyor belt for charcoal

Figure 38. The Reichert retort process.   (After: F.Flügge, Die Chemische Technologie D. Holzes)

-   Average yield: Charcoal 33-38 %; commercially obtained
    from 1 m^3 wood (390 kg dry wood): acetic acid (raw)
    22-23 kg, methanol (raw) 6-8 kg, tar and oils 53 kg.
    Energy demand: 0.7 million kcal/ton of dry wood;
    100 KWh/ton of dry wood. These figures do not include
    the electricity consumption of the sawmill (wood
    preparation).

-   Economic considerations: A battery of six retorts is
    necessary to sustain an economic operation. Complete
    control of the temperature during the carbonization
    phase guarantees high-quality lump charcoal. High
    investment and operational cost can be set-off by
    by-product recovery and the installation of automatic
    equipment. Expenses for retort maintenance are
    tolerable. The plant requires permanent supervision and
    well trained personnel.

3.1.2.3 The French SIFIC Process

The SIFIC can be considered as the most succesful
technology.

Figure 39 illustrates the scheme and modus operandi. The
importance (3, 5) of continuous carbonization of woodlogs
and sawmill wood waste can be seen in the increased turn-out
of products in relation to the investment capital costs, and
there is a definite saving in fuel costs for maintaining the
retort temperature.

The predried wood enters the top of the retort through a
lock. the level of raw material within the retort is
permanently kept at full load which can be electronically
controlled. During the charcoaling, the wood moves slowly
down towards the bottom. When passing the hot temperature
zone in the centre of the retort, it decomposes into
charcoal and gas. The gases are drawn upwards by a fan.

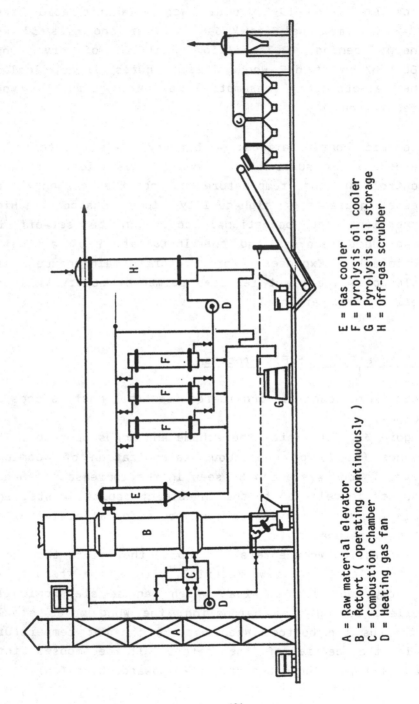

A = Raw material elevator
B = Retort (operating continuously)
C = Combustion chamber
D = Heating gas fan

E = Gas cooler
F = Pyrolysis oil cooler
G = Pyrolysis oil storage
H = Off-gas scrubber

Figure 39. The French SIFIC retort process.

The charcoal is discharged at the retort plenum and carried away to the storage. A separate charcoal cooler is not provided because the lower retort segment assumes this role.

In Figure 40, small auto-controlled lorries take the char to the desired locations to save manpower costs.

The SIFIC process can be run either without by-product collection, or with the necessary attachments for recovery (as explained in Chapter 4), the equipment and apparatus being the same. Their capacities, sizes, etc. must be adapted individually.

The heating system for the retort is commercially proved and functions according to the principle described in Chapter 3.1.2.2 (Reichert Retort Process). The fuel for the burning chamber is picked up at the scrubber for the necessary recirculating gas stream which passes through the retort charge.

Attention has to be given to the influence of wood moisture content and the energy demand of the heating system. The data in the following table have been obtained from commercial plants by carbonization of European hardwood species. The unit on which the energy consumption, the recirculating gas quantities and electricity were based is expressed in cubic metre of solid air-dry wood substance (fm). This is a common term in the local charcoal industry and 1 fm corresponds to 750 kg on average (hardwood only):

Moisture			Per fm (750 kg) necessary	
(%)	kcal	kj	Recirculating gas (m^3)	KWh
5	6.000	35.080	210	2.5
10	9.500	39.700	270	3.2
15	37.000	154.700	490	4.4
20	70.000	292.600	770	5.7
25	110.000	459.000	1.050	7.2
30	155.000	647.900	1.400	9.0

The proportionality of energy demand to raw material moisture is not the only problem. With increasing evaporation rates, the installed capacity of the plant is reduced and consequently production slows down.

These drawbacks can be eased or avoided if the incoming wood is passed through a dryer before entering the retort. The options for raw material drying and the details are described in Chapter 5.

Several attempts have been made to simplify the SIFIC process. One of the results was the CISR Lambiotte Retort. Plants have been commercially run for a number of years, and the main features are shown in Photo 12 and Figure 40. The retort is heated by an internal combustion device which burns part of the recycled pyroligneous vapours (off-gas). This compensates for the energy needs of wood drying, charcoal cooling and general heat losses of the retort.

The excess of pyroligneous vapours is taken from the top of the retort and is led back for use or re-use. To be easily combustible, the vapours must be produced from wood with a maximum moisture content of 30 %. The combustion air can be well controlled, thus ensuring oxygen deficiency and regulating the retort temperature.

The carbonization temperature is fixed to obtain a good quality of charcoal. In the lower part of the retort, a second gas circulation cools the charcoal before it leaves the converter. This gas stream is refrigerated and washed in a scrubber.

Operational data   (Lambiotte et Cie, S.A.)

-   Size of retort: height 18 m, diameter 3 m.

-   Possible throughput: approx. 7.00 tons of dry wood.

-   Raw material: wood logs, slabs, length 35 cm, diameter 10 cm (max.), moisture not exceeding 30 %.

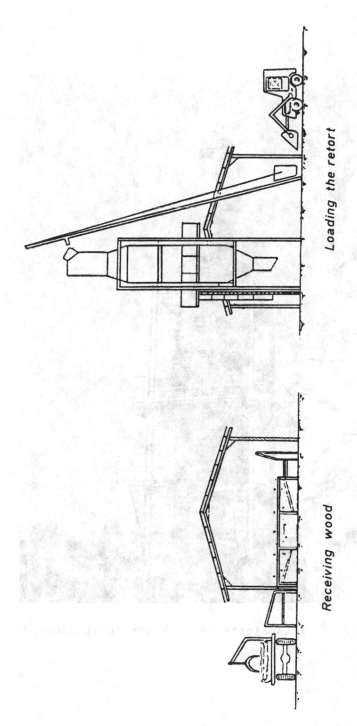

Receiving wood

Loading the retort

Figure 40 Side view of the CISR Lambiotte Plant

Photo 12   A Lambiotte reactor (Photo Lambiotte)

- Expected yields: approx. 2.500 tons lump charcoal;
  production of pyrolysis oil is feasible, but extra
  equipment is necessary.

- Energy requirements: for retort heating; electricity
  25 KWh

3.1.3 Charcoal Technology for the Carbonization of Biomass

The technology described in the preceding sections is
appropriate for the carbonization of stere wood or wood logs.
The following is a description of charcoaling equipment
designed for the processing of raw material which is
naturally composed of small particles, e.g. sawdust,
nutshells, or is deliberately crushed, chopped and reduced
to fragments, such as sugarcane bagasse, bark, twiglets,
coconut shells, etc. (4).

Because of the small diameter of these particles, the
heat exchange and transfer of energy from the converter
walls or heating gases through the surface of the small
particles is ideally fast and carbonization time is very
short. Therefore, this type of charcoaling is also referred
to as the "rapid pyrolysis process".

The first rapid pyrolysis process plant was the
previously mentioned Badger-Stafford process (see 3.1.1).
Since then many improvements have been achieved and the
systems have been made energy efficient. For more than
thirty years, the charcoal industry has employed multiple
hearth furnaces for this purpose. Because of the high
capital outlay for a multiple hearth furnace, experimental
development was initiated with small units built in the
early years of this technology.

The small-scale rapid pyrolysis technology has a
particular importance for the charcoal-maker in developing
countries:

- Developing countries, in contrast to the industrialized countries, abound in forestal and agricultural waste which constitutes a valuable raw material source for charcoal and by-product generation.

- Developing countries need to recover energy from indigenous sources to make them independent of expensive imports of fuel oil.

Almost all rapid pyrolysis plants can be designed for the generation of three products: solid char, pyrolysis oil, and converter gas (5).

The biomass char is obtained as coarse or fine powder and has to be agglomerated or briquetted in most applications. The cost of these additional operations is frequently overestimated. Moreover, the fact that briquetting improves the quality of the char and its value as fuel is not yet sufficiently well known. Briquetting and agglomeration technology will be discussed in Chapter 8.

3.1.3.1 Generalized Flow Diagram

Rapid pyrolysis plants are built according to a scheme which has been adopted for more than two decades. They can be differentiated by the retort/converter type only. Normally, a briquetting plant is attached (this is not shown in Figure 41).

The raw material is received, reduced in size by hogging (if necessary) and stored. Additional unit operations such as ferrous metal removal, glass removal, etc., may be performed before storing the sized material. From storage, the feed is retrieved and conveyed at a metered rate to the dryer. Here the moisture content of the feed is reduced to less than 10 % (wet basis), and the dried feed is then conveyed to a surge bin.

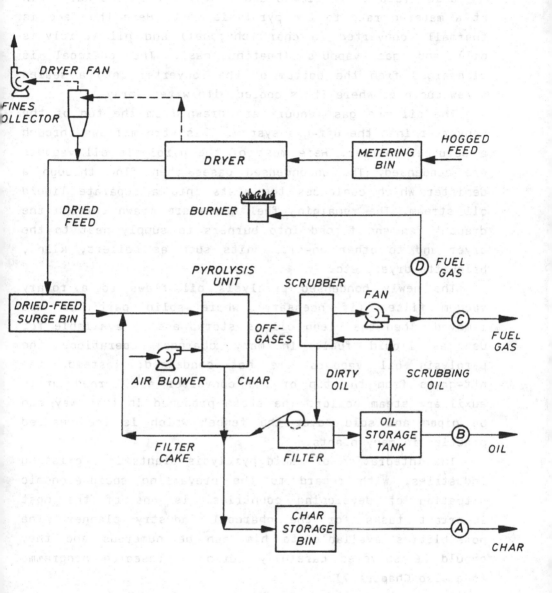

Figure 41   Generalized Flow Diagram of the rapid pyrolysis process.

(Courtesy of Carbon International, Ltd.
Neu-Isenburg, FRG)

Dried feed is retrieved and conveyed from the surge bin at a metered rate to the pyrolysis unit. Here the feed is thermally converted to char (charcoal) and oil (pyrolysis oil) and gas vapours (heating gas). The charcoal is discharged from the bottom of the converter into a sealed screw conveyer where it is cooled with water spray.

The oil and gas vapours are drawn from the top of the converter into the off-gas system. This stream flows through a scrubber-chiller. Here most of the pyrolysis oil vapours are condensed. The uncondensed gases then flow through a demister which coalesces the mists into a separate liquid oil stream. The remaining fuel gases are drawn through the draught fan and forced into burners to supply heat to the dryer and to other on-site units such as boilers, kilns, briquette dryer, etc.

The newly condensed pyrolysis oil flows to a rotary vacuum filter (if neessary) where solid particles are removed. Then the clean oil is stored and is available for use as liquid fuel. In many charcoal operations the pyrolysis oil vapours are not condensed; instead, the off-gases from the top of the converter are burned in an auxiliary steam boiler. The steam produced in this way can be piped and sold "over the fence" which is the desired practice in many plants.

The integration of rapid pyrolysis plants into existing industries, with regard to the prevailing socio-economic situation of developing countries, is one of the most important tasks for the charcoal industry planner. The possiblities available to him can be numerous and they should be surveyed carefully during a research programme (see also Chapter 7).

The criteria for rapid pyrolysis plants may be summarized as follows:

-    Conversion of small material fragments only.

- Limiting the moisture content of the raw material by mixing of dry with wet material, or pre-drying.

- Movement of the raw material during carbonization by free flow, forced flow or rotation.

- Short residence time of the raw material within the retort or converter to prevent product losses.

- Internal heating (type A, see section 1.3) whenever possible in order to simplify the operation.

- Control of product quantity and yield by varying residence time and carbonization temperature (process flexibility).

- Low capital investment within the means of small-scale producers.

- Substitution of automatic equipment and devices by labour if maximum of job opening is desirable.

3.1.3.2 The Multiple Hearth Furnace (6)

This furnace can best be described as a vertical kiln with a number of circular hearths which are enclosed in a refractory-lined steel shell (see Figure 42).

In the centre of the cylindrical shell, a vertical rotating shaft with radial arms moves the feed from the top hearth by rabbing teeth in a spiral path across each hearth. The material is thus constantly agitated and exposed by the rabbing teeth before it falls through drop holes from level to level.

The process air can be supplied in controlled quantities by combustion air blowers through burners or ports, or by induction through air ports. Automatic draught and

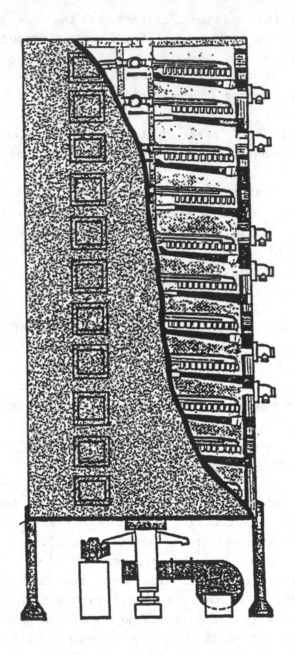

Figure 42. Cross-section of a multiple hearth furnace.
(Courtesy of ENVIROTECH, BSP DIVISION, Belmont, California)

oxygen-monitoring equipment are utilized to minimize power draw and fuel demand while ensuring clean combustion.

The shaft arms are cooled by forced air from a fan. The shaft has double-wall construction. Thus the cool air blows up the centre tube, through the rabble arms, and back into the shaft annular space to exit at the top of the furnace. Heated air is ducted back into the furnace for re-use as combustion air. Furnace refractory and layers of insulation in the thick walls conserve heat and maintain low shell temperatures.

There are several types of multiple hearth furnaces on the market. However, one of the most important points to bear in mind is that the furnace to be selected should have flexibility so as to handle a wide variety of materials of varying physical and chemical characteristics. Another point of interest to the potential charcoal-maker is the ability to utilize different fuels, including coal dust as well as waste oils.

The trend of technical development is to improve the overall economics of the multiple hearth furnace by adding equipment to make use of the converter gas. Therefore, systems have been developed with predrying installations and steam boilers to recover the energy contained in the remaining driven-off gases.

Another potential which should be considered by the charcoal production planner is the possiblity to install additional steam nozzles for charcoal activation. Since activated carbon has become a highly-sought-after commodity in industrialized countries, it should be mentioned here that the largest quantities produced are made by multiple hearth furnace activation. The investment costs for the multiple hearth furnace are very high, when compared with retorts and small-scale converters. Therefore, these have to be offset by a high charcoal capacity and increased productivity. As a rule of thumb, the lower limits for economical operation would be 15.000 to 18.000 tons of charcoal per year.

For the planner in developing countries, it should also be pointed out that this technology is designed for reduced manpower, and therefore does not necessarily create an attractive number of job openings.

3.1.3.3 The Fluid Bed Carbonizer

In this process the raw material (K) is directed to a bed of hot glowing charcoal in a closed chamber (D) (see Figure 43). This is maintained in a turbulent state by introducing an oxygen-containing gas under pressure into the bed.

To obtain better results, the gas is heated before reaching the bed, but it may not be so heated. The gas and particles may be introduced together or separately, and the gas may enter the bed at one or more points.

The glowing charcoal quickly distils and gasifies the wood particles. The oxygen-containing gas and the evolved gases are present in such quantity that the charcoal and particle bed is maintained in a turbulent or "fluidized state" resulting in a uniformly high temperature throughout.

Charcoal is formed continuously in the process, and the excess may be removed from the top of the bed in any desired manner. It may be removed periodically, but it is preferable to withdraw it continuously, and this is done by means of an overflow pipe (F), level with the top of the bed.

A certain amount of the charcoal is burned by the oxygen introduced into the bed, and the heat of this reaction supplies a portion of the heat necessary for the operation of the unit.

The evolved gases, containing acids, alcohols and tars rise through the hot charcoal body, a portion of the tar mist being cracked by heat to form lower molecular weight hydrocarbons.

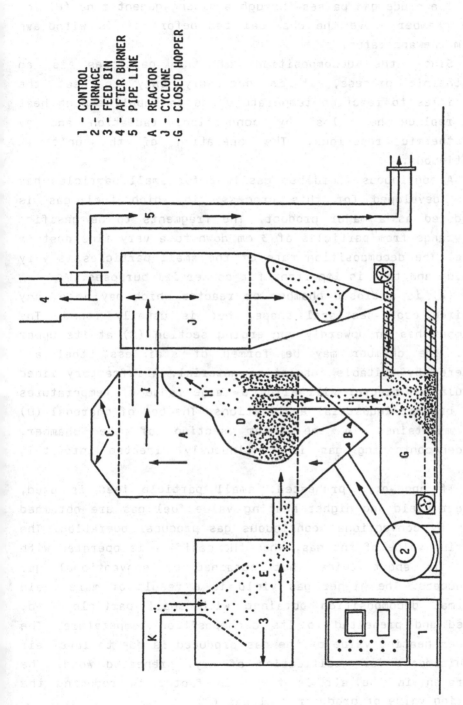

Figure 43. The fluid bed carbonizer. Generalized diagram.

1 – CONTROL
2 – FURNACE
3 – FEED BIN
4 – AFTER BURNER
5 – PIPE LINE

A – REACTOR
J – CYCLONE
G – CLOSED HOPPER

The crude gas passes through a disengagement zone (H) in the chamber above the charcoal bed before it is withdrawn from the apparatus.

Since the decomposition of the particles is an exothermic process, it is necessary only to heat the particles to reaction temperature and to supply enough heat to replace heat lost by conduction, radiation and by endothermic reactions. The operation of the unit is continuous.

A continuous fluid bed gasifier for small particles has been developed for this process, in which fuel gas is produced as a major product. The fragments to be gasified may range from particles of 3 cm down to a very fine dust in size. The decomposition rate of the small particles is very rapid, and this is important for commercial purposes.

(A) is a closed chamber or reactor which may be of any desired cross-sectional shape, but is usually round. The chamber has an upwardly converging section (C) at its upper end. The chamber may be formed of stainless steel and covered by suitable insulation material. A refractory lined insulation reactor will allow a large range of temperatures for broader industrial applications. The bed of charcoal (D) is maintained in the lower section of the chamber. Oxygen-containing gas is continuously directed into this bed.

If bone-dry, preheated, small particle feed is used, larger yield and higher heating value fuel gas are obtained than by conventional continuous gas producer operation. The heating value of the gas, when the gasifier is operated with air, is about twice that obtained by conventional gas producers. The higher gas yield is a result of more rapid thermal decomposition obtained with small particle feed, dried and preheated to its decomposition temperature. The higher heating value of the gas produced is due to lower air requirements for gasification of dry, preheated wood. The nitrogen in the air is the main factor in reducing the heating value of producer fuel gas (7).

Technical and Process Data (7)

Feed material: Sawdust, fruit pits, nutshells,
sugar-cane bagasse, max. size o.6 cm.

Achievable char yield: 18 % to 25 % of raw material (dry
basis) ca 2 m^3 gas per kg raw
material.

3.1.3.4 The Vertical Flow Converter (8)

The converter consists of an outside steel shell, a
composite refractory lining, an output feed mechanism at the
bottom, a lower plenum, a char screw, a char discharge
air-lock, an insulated top, and an air injection system. The
converter is normally run at a pressure slightly below
atmospheric and is sealed to prevent in-leakage of outside
air (see Figure 44 and Photo 13).

Dried feed material is fed through the air-lock near the
top of the converter; the flow rate through the converter is
controlled by the output system and a nuclear level detector
which controls the rate of feed into the converter.

As the material flows downwards through the converter,
it is heated to a maximum temperature of about 550° C and
thermally decomposes into charcoal (char) and gaseous
vapours. The char is discharged into the lower plenum where
it is water-spray cooled and conveyed through the screw and
air-lock to conveying equipment which takes it to storage.
The hot gaseous vapours flow upwards through the bed,
heating the downward-flowing solids, and exit through a port
at the top of the converter into the off-gas system.

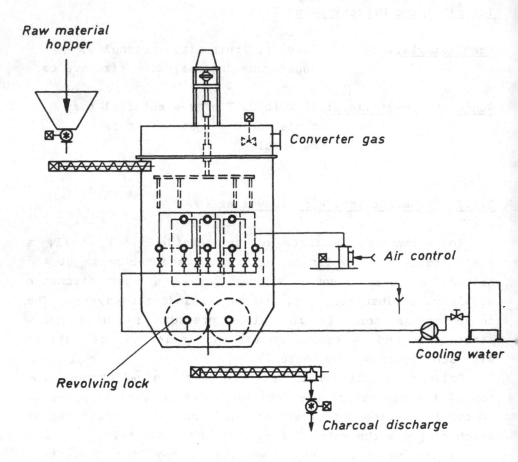

Raw material hopper

Converter gas

Air control

Cooling water

Revolving lock

Charcoal discharge

Figure 44   *The vertical flow converter.*

(Courtesy of AMERICAN CAN CO.
Greenwich, Conn.)

PHOTO 13 Model of a vertical flow converter charcoal plant.
Left: vertical flow reactor with gas pipes. Right:
raw material and charcoal storage.
(Photo Bio-Carbon, GmbH)

Raw material receipt

Raw material is received from a drag chain directly from a mill or by truck. The process feed conveyor incorporates a magnet for the removal of magnetic materials. The material passes over a screen which sorts out the oversized pieces and sends them to a hammermill (hog) for size reduction. The hogged material, together with the undersized material, is conveyed to a bucket elevator which loads the storage bin. From storage, a screw conveyor transports the sized material to a small metering bin to feed the dryer. The metering bin has level sensing devices which control the storage bin unloader in an on-off fashion.

Expensive equipment may be eliminated and manpower employed instead, wherever appropriate.

Raw material dryer

Material is fed at a metered rate from the metering bin to the dryer. The dryer can be one of several types:

- A single-pass rotary dryer

- A three-pass rotary dryer.

- A screw conveyer dryer of individual design.

In each of these dryers, hot gases are passed through the material. The inlet gas temperature varies with dryer design and ranges from 200° C to 800° C. The dryer effluent gas is passed through a cyclonic type separator to remove the entrained solid particles.

Solar or air predrying of the raw feed is recommended wherever applicable and will improve the overall economy of the process.

The off-gas system of the converter can be designed for collection of pyrolysis oil, including fractionation of the oil to produce raw material, raw acetic acid, creosotes and tar.

Technical and process data (8)

Converter: Forestal wastes, sawmill wastes, agricultural residues. Hogged to approx. size 5 x 1.5 cm.

Yield: 100 kg char per 350-390 kg of raw feed (dry basis). 80 kg pyrolysis oil per 350-390 kg of raw feed (dry basis).

Energy: 150 kcal/kg of dry feed; 0.05 KWh/kg of dry feed.

For the economic operation of a plant, a minimum throughput of 0.5 t/h of raw feed is necessary.

3.1.3.5 The Enerco Mobile Pyrolyser (Model 24) (14)

This system optimizes the quantity and quality of charcoal production. The Enerco method of recirculating the indirectly heated inert gas keeps products of combustion and nitrogen out of the pyrolysis reactor, thereby generating the highest grade of wood gas and distillates. Moreover this direct contact of the hot recirculated inert gas with the wood feed maximizes the rate of the char reaction.

The producer gas can be used directly in a user's boiler and/or condensed into a Bunker C type low sulphur oil. Heat obtained from complete incineration of unused and uncondensed off-gas sustains the char reaction. The waste heat both from the incinerated products of combustion and the condenser is made available under adequate gas pressure to ensure the proper operation of wood-drying equipment (see 5.3). The pressure varies widely depending on the type of dryer, material to be dried, and the drying rate.

The equipment is intended to operate outdoors in most weather conditions. Materials and workmanship are such that the apparatus can perform without interference from snow, rain, frost or freezing rain. Any components that cannot meet these requirements must be protected.

The maximum diagonal dimension of wood particles should not exceed 1.8 cm. The process is designed to handle a wide range of feeds, from unsized sawdust to hogged or chipped wood. There is no requirement to size-sort the feed material prior to pyrolysis. At a given pyrolysis temperature, the char time is a function of the dimension of the largest piece to be charred (always taking the smallest dimension of a given particle).

The wood feed is automatically loaded into the top of the reactor by a 23 cm auger which elevates the feed from floor level. Charcoal is unloaded on a presettable time/volume schedule by a 23 cm auger which elevates the charcoal to a height of 1.2 m (see Figure 45).

Both augers are tightly flanged to the reactor, thus preventing air or gas leaks at these points. A flange is provided at the exit end of the off-load auger for convenient sealing to the user's char-handling facility.

Moisture content limitations on the feed

Pyrolysis of wood feeds up to 23 % moisture by weight has been carried out successfully. The drier the feed, the quicker will it produce charcoal and the higher will be the energy value of the off-gas stream. To lower the moisture content of the feed, a heat exchanger is included with the pyrolyser to provide warm, temperature-controlled air to the infeed surge bin.

For each 3 tons of bone-dry wood infeed, the "24" produces 1 ton of char, 1 ton of gas, and approximately 1/2 ton of oil. The amounts will vary with type of feed (species, bark concentration, etc.) and pyrolysis temperature.

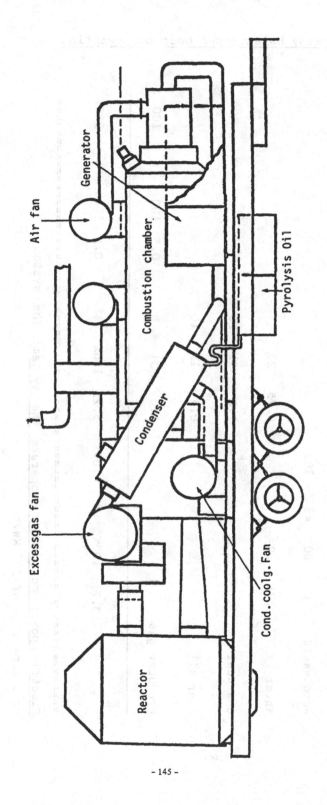

Figure 45. The ENERCO Model 24 Pyrolyser (right side).

Mass and energy balance per hour of operation

Material	Input				Output				Consumed in Process			
	Tons	%	Btu(10^6)	%	Tons	%	Btu(10^6)	%	Tons	%	Btu(10^6)	%
Wood waste	3.0	80	48	100								
Water	0.75	20			0.75	23						
Charcoal					1.0	31	26					
Char oil					0.5	15	10					
Producer gas					1.0	31	8		0.5	100	4	100
	3.75	100%	48	100%	3.25	100%	44		0.5	100%	4	100%

==

Electric power consumed to operate the system: The system uses a maximum of 30 KW/h.

References

(1)   M. Klar, Der Stand der Technologie und die Wirt-
        schaftlichkeit der Holzdestillationsprodukte,
        Wochenblatt für die Papierfabrikation, Nr. 5/6,
        1936.

(2)   H.M. Bunburry, The Destructive Distillation of
        Wood, London, 1925.

(3)   F. Fluegge, Chemische Technologie des Holzes,
        (36-71), Munich FRG, 1954.

(4)   W. Emrich, "The Carbonization of Biomass", paper
        presented at the Nuclear Research Center, Juelich,
        FRG, 1981.

(5)   W. Emrich, "Recovery of Char and Pyrolysis Oil from
        Forestal Waste and Agricultural Residues by
        Carbonization", paper presented at the
        International Recycling Congress, Berlin, FRG, 1982.

(6)   Information provided by Envirotech, Eimco BSP
        Division, Belmont, California, USA 1982

(7)   Information provided by Protran Co., Raleigh, North
        Carolina, USA 1982.

(8)   Information provided by American Can Co.,
        Greenwich, Conn., USA 1983.

(9)   Information provided by Enerco, Inc., Langhorne
        Pennsylvania, USA 1982.

Chapter 4
TECHNIQUES FOR RECOVERING COMMERCIAL PRODUCTS FROM PYROLYSIS OIL

The off-gas leaving the retort or converter is at a temperature of 280° to 300° C. It contains the pyrolysis oil vapours and the uncondensable residual gas which is often referred to as heating gas.

In many cases, this high energy containing off-gas is burned for the sole purpose of firing a steam boiler furnace. It has a tendency to clog gas pipes if not properly handled and if the pipes are not well insulated. Therefore, extensive pipelines must be avoided. These problems may be minimized by stripping and scrubbing it from the pyrolysis oil.

It has been mentioned already that pyrolysis oil was once an important raw material for the chemical and other industries. Today, although few charcoal plants make use of recovery technologies, the interest in this raw material source is increasing (1).

Factories specializing in carbonization/distillation processes usually confine themselves to delivering crude products (pyroligneous acid and tars) to specialized distilleries. The rectification and separation of methanol, acetone and acetic acid in the pure state that is required is a delicate process and has no direct connection with carbonization itself.

The products of the normal medium-sized charcoal plant are: wood spirit (methanol), acetic, propionic, and butyric acid, tar and creosotes.

The outlook for chemicals derived from pyrolysis oil cannot be forecast at this time. However, it seems worth while to acquaint the charcoal planner, particularly in developing countries, with the basics of this special field. There is a common belief in the charcoal industry that those developing countries which are building up their own

chemical production capacities will be more likely to fall
back upon these domestic resources.

Until now, the great number of raw materials available
to the charcoal industry in the tropics has been little, or
not at all explored from the viewpoint of recovery of
pyrolytic products. It remains to be seen wether the
governments and their agencies concerned with energy supply
will give more attention to this area of charcoal-making in
the future.

4.1 Pyrolysis Oil Recovery

The first step in pyrolysis oil devaporisation is
cooling of the retort gas. The surface area of the cooling
system depends in the first place on the quantity of raw
feed input in the retort or converter. Theoretically, 1.90
to 2.00 m^2 heat exchange area is necessary for 100 kg dry
feed, if water is the cooling agent. The resulting gas will
be at a temperature of 20-30° C and contain CO, hydrogen,
CO_2 and also traces of pyrolysis oil vapours.

In practice, far greater heat exchange areas must be
installed. The designer must have sufficient experience of
charcoal technology, because there is no way to calculate
the actual size of the pyrolysis oil cooler and test runs
are usually not possible.

It is advisable to protect the cooling system against
stray raw material particles by inserting a cyclone or
demister between it and the retort or converter. This is a
necessity for continuous carbonization, and rapid pyrolysis
plants in particular. Without it, clogging and congestion of
coolers takes place which cause interruptions of the
operation (2).

All parts of the equipment in this section of the
operation and the following stages, which come in contact
with the products must be made from corrosion-resistant
metals, usually stainless steel or copper.

4.2 Crude Acetic Acid and Acetone Recovery

The wood spirit is first eliminated in a steam-heated rectifier. The pyroligneous acid rises half-way up the rectifier column. The wood spirit distills off at the upper part whilst the pyroligneous residue runs off in the lower part. These two crude products are discussed in another section.

Lime acetate process

The pyroligneous acid, after the separation of crude wood spirit as above, is distilled in copper vessels and the water and acetic acid vapours are collected in a limewash while the tar in solution is left in the retort (retort tar).

The lime acetate so precipitated is dry-evaporated in drums at a temperature not exceeding 180° C. This crude acetate, known as grey acetate, contains 80 % by weight of calcium acetate.

The solubility of calcium acetate varies little with temperature; hence, it is necessary to carry out dry evaporation, but such evaporation ceases when the mass congeals into a crystalline magma containing only layers. To convert the magma into grey acetate at 80 per cent, evaporation must be carried out on shallow layers. It is impossible to obtain an anhydrous acetate, for at the temperature needed to evaporate the last 20 per cent of moisture the acetate itself would begin to decompose.

By decomposing grey acetate with sulphuric acid, an acetic acid at 85 per cent, contaminated by sulphur derivatives, should be obtained. The treatment of grey acetate with sulphuric acid should be carried out in heated mixing apparatus. Calcium sulphate and sulphuric acid are formed on distillation of crude acetic acid, later purified by rectification.

Acetone can be obtained from grey acetate by heating thin layers of the latter in mixing apparatus at 400° C to

500° C. The crude acetone is purified by rectification.
These processes are not more used in modern plants.

Direct recovery of acetic acid by distillation

The process starts from a weak acetic acid solution
obtained by the distillation of pyroligneous acid leaving a
residue of retort tar.

The process employed by several companies uses toluene
to draw off the moisture during the second distillation. The
process is based upon a principle development to determine
the moisture content of wood.

A more sophisticated but well designed two-stage process
is used by a French company. First, the solution is diluted
with xylene; on distillation, the latter combines with water
to form a binary mixture with very low boiling point. The
first-stage concentration of acetic acid is then distilled
again with benzene, forming a fresh mixture with a still
lower boiling point.

A process of far-reaching practical importance is that
used both by the Société des Produits Chimiques de Clamecy
and Lambiotte Brothers plant (3, 4). This method is also
based upon the elimination of moisture by distillation of an
azeotrope formed with butyl acetate, or with a fraction of
wood spirit passed into a tar extractor in which a
counter-current of butyl acetate circulates. The latter
serves to maintain a constant water vapour content in the
acetic acid vapour. This vapour is then passed through a
dehydration tank containing butyl acetate. At 90° C, an
azeotrope distils off at the upper part of the tank; the
azeotrope contains 71.3 per cent butyl acetate and 28.7 per
cent moisture by weight. The heating of the tank is
controlled so that condensation takes place at the bottom of
the tank. Two layers are deposited: the upper one,
consisting mainly of butyl acetate, returns to the tar
extractor; the heavier layer, consisting of a solution of 30
per cent acetic acid in water, is conveyed to a further tank

for concentration. The azeotrope distilled at the top of the dehydration tank is condensed and separates out into a light layer containing 98.9 per cent butyl acetate and a watery layer containing only 0.8 per cent of the latter. The light layer is returned through the dehydration tank (see Figure 46).

The plant belonging to the Deutsche Gold und Silber Scheidenanstalt uses the acetic ether method for the extraction of acetic acid. Pyroligneous acid freed from wood spirit is introduced into the top of a 10-metre tower filled with the ether. At the same time, a counter-current of acetic ether is introduced from the bottom. The tower is fitted with rashig rings to ensure good contact. Working is continuous (5).

At the bottom of the tower, there is a run-off of water containing some dissolved ether which is collected by rectifying. A mixture of ether and acetic acid with a little tar is given off from the top. The ether is collected by rectification and recycled into the extraction tower. The separation of crude acetic acid and tar is effected by vacuum distillation. The resultant acetic acid is at 95 per cent; it is rectified.

It is not known whether this very attractive process can offer the same guarantees of reliability as those using the various French techniques.

Acetone can be obtained by passing the pyroligneous acid (after separating out wood spirit) or the pyrolgneous distillate over catalysers at 500° C.

4.3 Recovery of Methanol (Wood Spirit)

Treatment of wood spirit

The wood spirit separated as above from the crude pyrolysis oil contains about (6, 7):

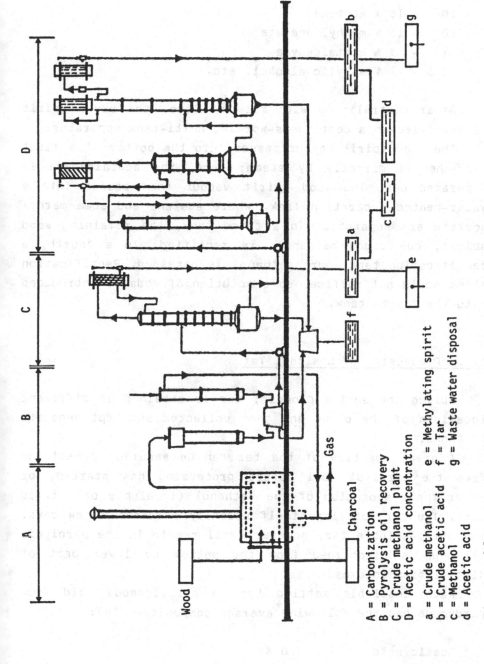

A = Carbonization
B = Pyrolysis oil recovery
C = Crude methanol plant
D = Acetic acid concentration

a = Crude methanol   e = Methylating spirit
b = Crude acetic acid   f = Tar
c = Methanol   g = Waste water disposal
d = Acetic acid

Figure 46. A charcoal plant with pyrolysis oil refinery.

```
65   - 70 % methanol
10   - 15 % acetone
10   - 15 % methyl acetate
 1   -  3 % acetaldehyde
0.5  -  1 % allylic alcohol, etc.
```

After neutralizing with milk of lime, the crude spirit is rectified in a continuous-working multi-tank apparatus.

The wood spirit vapour passes into the bottom of a first tank heated directly by steam; here, the acetaldehyde is separated out. The wood spirit vapour then passes into a water-heated separation tank, where acetone and some methyl acetate are separated. In a third tank, the remaining wood spirit, now crude methanol, is rectified. In a fourth, a rectification tank, pure methanol is obtained. Rectification alone would not suffice, so a solution of soda is introduced into the fourth tank.

4.4. Processing of Charcoal Tar

During the entire process, tar is obtained at different locations of the plant and then collected and kept separate in tanks.

A great quantity of the tar can be secured by settling from the pyrolysis oil before processing has started, or after the evaporation of the methanol (insoluble or settled tar). The tar settling itself normally requires a few days. The water-soluble tar, however, will remain in the pyrolysis oil and is discharged from the bottom or lower part of distillation columns.

Water-insoluble settled tar in pyroligneous acid from hardwoods has the following average composition (8):

```
Acetic acid        2.0 %
Methanol           0.7 %
Moisture          17.7 %
```

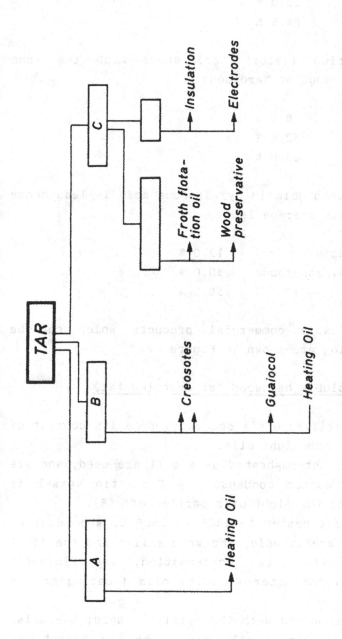

A = Tar oil, light
B = Tar oil, heavy
C = Tar residues from still

Figure 47 Recovery of commercial products from pyrolytic tar.

Light oils	5.0 %
Heavy oils	10.0 %
Pitch	64.6 %

Tar in solution (retort tar) shows much the same composition in softwood or hardwoods:

Acetic acid	8.0 %
Moisture	32.0 %
Pitch	60.0 %

Softwood tar insoluble in pyrolgneous acid is less dense than the latter. Its average is:

Pyrolytic acids	12.0 %
Terpenes and hydrocarbons	30.0 %
Tar proper	58.0 %

The most valuable commercial products which can be gained by processing are shown in Figure 47.

Processing of insoluble hardwood tar (settled tar)

The tar is distilled until separated from its content of water, wood spirit and light oils.

Copper retorts, steam-heated by a coil are used, and are connected with a copper condenser. A Florentin vessel is used for separating the light oils carried off (8).

The tar is first heated to 140° to 150° C, when most of the moisture, the acetic acid, the wood spirit and the light oils are given off. After condensation, the Florentin receiver separates the water from the oils floating on the surface.

The water is reunited with the pyrolytic acid; the oils, which have an unpleasant smell, are of no use except for heating.

The tar is heated further with the direct introduction

of steam until no more acetic acid is carried off. The retort residue is then a tar free of moisture and acid but still containing heavy oils. This process neither rectifies the heavy oils nor separates the creosote; it simply enables the acetic acid in the tar to be collected.

To separate the creosote, the tar should be distilled in an open fire retort similar to those used in distilling coal tar, but the top should be made of copper to resist the acidity of the volatile substances. The base of the retort is of cast iron.

The preliminary heating should be done with care, for the tar tends to froth. The pyroligneous acid and the wood spirit distil first, issuing from the condenser as a yellowish, watery liquid with light oils floating on top. The first stage of distilling is completed around 110° C, and the temperature should be raised to 250° to 260° C before distillation is resumed. The oils which are respectively lighter and heavier than water are collected in different vessels. These crude oils are not yet saleable for they contain appreciable amounts of tar.

The operation is completed at 260° C and pitch remains.

Oils from tars have lost their value with the rise of petroleum products; heavy oils are still somewhat appreciated for their creosotes which are used for wood preservation. These heavy oils are washed with a solution of common soda, the acetic acid of which combines with the pyrolytic acid. The resulting purified heavy oil contains 50 per cent phenols, and creosote, and is often known as "oil of creosote".

The creosote can be extracted from it by washing with a weak soda solution which transforms all phenols and the creosote to salts. The separated alkaline solution is steam-injected to remove the traces of oil remaining at the bottom.

The saline solution is treated with a dilute mineral acid, and the crude creosote rises to the surface. This creosote is not pure, but the guaiacoal can be separated from it by successive treatment with soda, steam and acid and lastly by distillation.

Creosote used for wood preservation is really oil of creosote.

Processing of insoluble softwood tar

The composition of this tar depends to some degree upon the charcoal-making process used; it is worth while to return to this point.

When normal softwood (20 per cent moisture) is distilled, the moisture begins to volatilize, carrying off turpentine. As the temperature rises in the retort, the resin grows soft and gathers on the face of the wood. At 180° C, the resin distills, and the products given off are, on the first cracking, the "pine-oils" followed by resin oils at increasing boiling points.

When large amounts of wood are distilled, an evenly distributed temperature is impossible. The turpentine carried off by the water vapour from the less heated areas of the retort is accompanied by products resulting from resin breakdown emanating from the hotter areas. Hence the turpentine so obtained is mixed with pine oil. Sometimes there is a mixture of tars from still hotter portions of the retort. This crude turpentine is called "Crude pine oil" or "German spirit" (9).

The smaller the distilling apparatus, the more constant is the composition of the distilled products, owing to evenness of temperatures. As a rule, the distillation of softwoods on an industrial scale produces turpentine which condenses only in the coolers and tar which condenses between the retort and the condensers properly. This tar contains some pine oil.

The crude turpentine is given off with steam. Turpentine, crude commercial pine oil, and a commercial tar are separated. The crude commercial tar is refined to extract oil of turpentine by heating it to 130° to 150° C and injecting steam which carries off all products except the tar. The resulting distillate is washed with soda solution which fixes any fatty acids and the phenols, as well as resinifying aldehydes and condensing acetone. The alkaline solution is separated and treated with dilute sulphuric acid, so eliminating the furanes, the aldehydes and the unsaturated compounds.

The washed oil is rectified either with an uncovered furnace or in a steam bath, the quality of the product being improved by vacuum distillation at 110° to 120° C. The resulting distillate is a commercial oil of turpentine.

Treatment of soluble tar (extraction tar)

On being distilled, retort tar gives only two kinds of products: a pyroligneous acid comparatively rich in acetic acid, and a very brittle pitch. Oil fractioning is non-existent. This tar has no value in itself, but its acetic acid content is collected either by open-fire distillation or (better) by steam treatment.

The tar is loaded into a copper retort, with both open and closed heating coils. The closed coils raise the temperature to distillation level, and then the open coils enable collection by steam. The operation continues until the acetic acid is practically exhausted. The pitch is removed before cooling. At ordinary temperatures it becomes brittle and can be used as fuel.

4.5 Concluding Remarks

It is obvious that the overall energy balance of a charcoal plant depends on the efficiency of by-product recovery.

However, the overall economic balance depends on the effectiveness of the commercialization of the recovered by-products in current markets.

By employing current technology and advanced equipment, a modern charcoal plant can be operated as an energy self-sufficient entity (10).

Industry has reported energy utilization factors, as compared with the total energy input, of up to 85 %.

It is commonly believed in the industry that the energy losses of charcoal plants cannot be reduced below 17 %, perhaps even 20 %. These "natural" losses result from the escape of tangible heat through the insulation of equipment, through cooling water, and heat losses from the freshly-discharged charcoal which is well above ambient temperature.

Evidently, the investment of a potential charcoal-maker is limited by cost-benefit considerations, in particular by the market value of his products. Consequently, many charcoal-makers have no choice but to accept a global energy balance well below the optimum of 85 % as a result of lack of demand.

Looking to the future, it should be noted that several institutes and private companies have resumed their research programmes on the optimal utilization of charcoal by-products. This research line had been neglected during the period of cheap petroleum and natural gas.

This branch of specialized chemistry dealing with the modification of pyrolysis products still encounters difficulties inherent in the nature of pyrolytic compounds, for example the strong tendency to self-polymerization of some of these products. However, research in this difficult area is progressing; in particular, there is significant progress in the catalytic heat pressure treatment of pyrolysis oil.

While awaiting the results which are expected from the research efforts, it must be accepted that, for the time being, pyrolytic oils and charcoal gas will be restricted to the replacement of fuels for steam boilers, brick factories, cement and fertilizer plants, gas engines, etc.

References

(1)   Chemicals from Wood are Economical Now, Chemical
        Engineering News, Dec. 6, 1976

(2)   M. Klar, Heutiger Stand der Technik und Wirtschaft
        der Holzdestillationsindustrie, Wochenblatt für
        Papierfabrikation, Nr. 5/6, 1936

(3)   French Patent No. 696803, Soc. des Produits
        Chimiques de Clamecy, 1929

(4)   French Patent No. 760593, Etablissements Lambiotte
        Frères, 1933

(5)   German Patent No. 592119, Degussa, Frankfurt, 1930

(6)   G. Petroff, Pyrolyse des Bois Tropicaux, Influence
        de la Composition Chimique des Bois sur les
        Produits de Distillation, Revue Bois et Forêts des
        Tropiques, no. 177, 1978

(7)   J. Doat, La Carbonisation des Bois Tropicaux, Revue
        Bois et Forêts des Tropiques, no. 159, 1975

(8)   W. Sandermann, Die chemische Verwertung von
        Stubben, Markblaetter des Reichsinstitutes fuer
        Forst- und Holzwirtschaft, 1948

(9)   D. Zinkel, Chemicals from Trees, Forest Products
        Laboratory, Madison, USA, 1975

(10)   W. Emrich, Sierra Leone. Production of Charcoal
        Briquettes and Pyrolysis Oil from Agricultural
        Wastes, UNIDO, Technical Report, 1981.

Chapter 5
RAW MATERIALS SUPPLY

Until two decades ago, roundwood or pile wood was the primary raw material for charcoal. Because of the competition by other industries, particle board and fibre board manufacturing wood prices became unaffordable for the charcoal industry in European countries and the U.S.A. Thus the charcoal-maker turned to other sources like slabs and off-cuts of lumberyards and timber sawmilling, bark and sawdust.

Yet today, most charcoal is still made from wood extracted out of natural forests and generally about five tons of wood produce one ton of charcoal. For every person in a community who uses charcoal for heating and cooking, about 0.5 ha of high forest has to be set aside to provide that wood supply in perpetuity, if one follows through official statistics (1).

Therefore, charcoal making can only be an on-going industry where improved and advanced technology is employed and the raw material resource is managed to provide a continuing supply.

Looking at the encouraging progress which has been made in Argentine and Brazil one can easily recognize that only a tenth of the area above would be adequate if the wood comes from well managed fuelwood plantations and is more efficiently used. Certainly, the use of wood derived from sawmill waste and land-clearing residues for charcoal-making does in no way alter the long-term forest land or plantation requirments for fuelwood.

The two major ways to achieve the objectives of resource management of the raw material supply for charcoal-making are to make the forest more productive by improving growth and reducing waste in harvesting, and to improve the conversion ratio of the raw fuelwood to finished charcoal and its by-products (2).

A natural forest is a resource which, in the economist's jargon, grew without labour inputs from man. The aim of forest management is to harvest a maximum timber crop from such a forest without destroying its productivity as an on-going ecosystem and, at the same time, minimize the inputs needed to achieve this. The result of this process is expressed in the mean annual allowable offtake or cut of the forest, usually measured in cubic metres per hectare. Theoretically, one could remove a volume equal to this each year and the forest would maintain itself. In practice, the intervention of man produces long-term changes in the forest, especially in the tropics, changing the species composition and the diameter classes of the mature, natural forest after harvesting and regeneration. Wherever possible, a forest should be managed to produce the product mix of highest value - sawlogs and veneer logs are first priority. Fuelwood has the lowest value; it is wood which cannot normally be sold for any other purpose. Its price is usually below pulpwood for the paper industry.

The usual compromise achieved - even in countries where forest management is strong and well-oriented - is that a certain area of forest is allocated for fuelwood supply with the annual allowable cut set at a level believed sustainable from the knowledge available at the time. The fuelwood harvesting enterprise then endeavours to stay within the prescribed cut and to maximize the harvest by making effective use of branchwood, dead timber and small diameter wood of poor quality, etc, which is not normally included in the assessed standing volume for yield calculations. To avoid damaging the forest system, however, there needs to be constant monitoring and measurement by the forest management authorities to ensure that target regeneration and growth rates are being achieved and to decide whether the allowable cut may be increased or must be reduced.

As already mentioned, other sources which are ideally suited for charcoal-making are the residues of forestry and the primary and secondary wood industry (3).

Numerous feedstocks have been tried out successfully by carbonization during the last three decades. Among those which have found widespread commercial use are:

- leftovers from clear-cuttings: brushes, branches, leaves, roots

- waste of lumber mills and the furniture industry: sawdust, off-cuts, slabs and wood shavings

- residues of the pulp and paper industry, bark

As regards the yields of charcoal, pyrolysis oil and heating gas, no significant differences exist between pile wood and these residues.

In general, it does not make much difference whether hard- or softwoods are utilized if the charcoal is used as a cooking fuel only (4). Softwoods will produce a lighter charcoal which also tends to higher attrition and abrasion rates. This makes it undesirable for some industrial applications as reductant for blast furnaces and activated carbon. The resins contained in conifers have slight deleterious side-effects, such as an unpleasant flavour imparted to barbequed food.

Many methods and concepts have been tested to balance out the lower density of softwood charcoal by alteration of the carbonization process, but with very little or no effect. The most promising way to make an adequate industrial charcoal from softwoods is by agglomeration with special binders which have the capability to reinforce the formed char.

The dissimilarities between pyrolytic oils made from hardwood or softwood are more significant. In general, softwood pyrolysis oil is higher in calorific value, presumably due to the resin content in the wood.

Furthermore, extremely good yields of tar and naval stores (turpentine, pine oil) can be extracted from these oils, which makes them a valuable fuel (5).

5.1 Supply from Fuelwood Plantations

Producing wood for charcoal from plantations demands that the cost of producing the fuelwood on the stump be carefully calculated to ensure that such a long-term investment is, in fact, worth while. On the other hand, the cost or stumpage of wood from natural forests is arbitrary and is set, in effect, by ordinary market forces, somewhere between zero cost where a small-scale charcoal producer gathers wood without payment from vacant forested land, and the cost of producing equivalent fuelwood from plantations. State forest services sometimes attempt to set fuelwood stumpage by calculating the management cost of the natural forest from which the wood is taken. Sometimes private natural forest owners set a stumpage rate as a percentage of the value of the charcoal produce. Around 10 per cent is a typical charge. Government stumpages are usually less than this when expressed on the same basis (6).

Given the problems of fuelwood and charcoal supply in many developing countries where natural forests have been cleared, or otherwise devastated, forest science has developed systems for cultivating man-made plantations of quick growing forest trees. The eucalyptuses native to Australia have been widely adopted and modified by selection for this purpose throughout the world. FAO's book "Eucalyptus for Planting" provides a wealth of information in this field and is essential for anyone seriously interested in this area.

There are many species of eucalyptus used in plantations, allowing adaptation to particular local conditions, and fortunately all make excellent fuelwood and charcoal. Where plantations are established and managed correctly on suitable sites, growth can be rapid. Mean annual increments (MAI) of 15 - 20 m^3 per ha over 12 - 20 years rotations are not uncommon.

The establishment and management of fuelwood plantations is a specialized branch of forestry. Further discussion would exceed the scope of this handbook.

5.2 Agricultural Resources

With the development of small-scale equipment for biomass conversion, the vast area of agricultural waste has become a potential source of charcoal. In this sense, wastes mean the leftovers from plantation harvesting or the discharge from processors of agricultural crops (7).

Experimental development and pilot-plant tests with these materials started before the Second World War. Commercial exploitation of the research results began only in 1950 in the U.S.A. The list of the agricultural residues which have been successfully tried in the meantime comprises more than forty different raw materials. They can be grouped as follows:

- husks and shells of nuts

- plantation residues from coffee, cotton, papaya and orchards

- discharges from farm cropping and food processing

- sugarcane bagasse

- straw, reed, bamboo, grass, weeds, underbrush, cactus

- assorted municipal waste

- industrial waste from the carpet industry, pulp and paper, slaughterhouses.

The quality of charcoal made from different agricultural residues can vary greatly, e.g., nutshells are a classic raw material for activated charcoal. The char displays a high degree of hardness with low attrition rates.

The natural mineral content of some raw materials severely limits their applications, because all the minerals will be found eventually in the char and increase its ash content. Consequently, the calorific value of charcoal made from such materials is poor, and industrial consumers requiring low ash content would reject it. Therefore, high ash containing charcoal is usually briquetted and sold as cooking fuel.

The yields of charcoal made from agricultural residues also vary greatly and cannot be compared to those made from wood or wood waste. Yields range between 15 % and 45 %, based on the dry feed. The same variations can be observed in the output of pyrolysis oils. Also the composition and burning characteristics of these pyrolysis oils are very different.

Research in this field has not yet explored all the opportunities for industrial uses. Investigations are going on to test the oil as a source of alternative car fuel. For the time being, its major commercial value remains as a substitute for fuel oil in industrial boiler furnaces.

One of the advantages of agricultural residues for the charcoal plant is that large quantities can be secured in almost bone-dry condition. A supply of this kind makes predrying equipment obsolete and reduces operational costs considerably.

It remains to be seen what efforts developing countries will make in this field with a view to reducing their energy bills and securing cheap cooking fuel for the people. Most of these countries abound with untapped agricultural waste resources.

Before the introduction of small conversion units, the economy of the charcoal plant depended on the concentration of, and the cost of collecting huge raw material reserves. The picture has changed completely: already, relatively small waste accumulations will sustain a charcoal plant economically.

5.3 Transport and Preparation of Raw Materials

Getting the fuelwood from the tree in the forest to the plant site can be very costly. A guideline of the charcoal industry is to keep the transportation distance of the fuelwood a minimum, and instead, carry the less bulky charcoal the longer distance. How close one can come to the ideal situation depends very much on the charcoal technology employed.

There is always a trade-off between the fuelwood transport distance and the cost/yield ratio of the carbonization process. At one end of the scale, there are the pit and the portable metal kiln technologies which need a minimum harvest transport distance. At the other end of the scale are the technologically complex, capital-intensive, large scale retorts and the multiple hearth furnace systems which are fixed installations: these imply relatively long transport distances for fuelwood. Brick kilns having a life of several years imply an intermediate distance for fuelwood transport. The fuelwood transport distance associated with brick kilns and high technology retorts and furnaces depends on the fuelwood yield of the forest and the expected life of the equipment for carbonization. Retorts which may last thirty years or more require a large block of forest so that they can be supplied with wood at the minimum haul distance during their useful life. Brick kilns having a life of about five years require sufficient forest to maintain fuelwood supply for this period before increased transport costs force the kilns to be moved to a new area.

In other words, the higher the yield/input ratio and the more sophisticated the equipment to be operated, the less important does the transport distance become.

In practice, the small charcoal-maker who works with his charcoal pit or simple kiln method will depend totally on one type of raw material and must move as close as possible to his resources. The charcoal-maker operating a biomass conversion plant may choose his input among many feedstocks, including trees from the forests, and distances to his raw material resources become of less concern to him.

The preponderant majority of biomass plants are located at the source of the waste discharge. They generate significant quantities of energy for supply to an existing sugar factory, etc. Therefore, they are sometimes attached to such facilities which can offer more convenient operating conditions and better maintenance.

Several waste materials are naturally in a state suitable for immediate consumption, e.g., rice husks, coffee husks, nutshells, cotton bin waste, etc. Others will require predrying and compaction: sugarcane bagasse, vetiver grass, reed, etc. For the conversion of sawdust, only a predryer will be necessary, whereas bark usually makes additional shredding unavoidable.

These few examples are given as a guide to the prospective charcoal-maker.

5.3.1 Key Factors in Wood Supply

Harvesting and transport may be analysed by breaking the process down into "unit operations" and treating these units as cost centres to determine their influence on total costs. The "unit operations" in harvesting are:

- Roading the forest compartment and defining the coupes or harvesting units of the compartment.

- Felling and bucking to required lengths; splitting may be required.

- Primary transport to secondary collection point.

- Drying of fuelwood in the forest.

- Secondary transport to the carbonization unit.

- Drying and storage of wood at the charcoal-making centre.

The above processes may be further subdivided, or some operations may be combined and others omitted in particular cases.

In the above unit operations, the only two which are significantly influenced by the distance between the charcoal production centre and the logging site are the primary and secondary transport of fuelwood. In the case of fully portable systems, i.e. pits, earthmounds and metal kilns, secondary transport is eliminated and primary transport remains more or less constant. For brick kilns it is different. Primary transport can be held constant, if desired, by laying out the forest area with a closely spaced access road network which reduces primary transport to a minimum.

In the case of an industrial charcoal plant, the transport costs have to be borne fully and are dependent on the mileage of the trucks, which increases every day. However, in practice many charcoal plants have been located close to or within an existing lumber yard, sawmill or furniture manufacturing plant.

Transport costs usually include the payment of labour to load and unload the truck, rail wagon or barge. Many attempts have been made to simplify this work and to save labour costs. One of the most successful ways is to stack the pile wood on wire ropes. When lifted by a crane, the pile wood forms a bundle which can be carried over to the transport vehicle. Each bundle contains four to five steres, thus considerably shortening the stowing and discharging time of the cargo. In all known plants using this procedure, the work is being done by the driver and his mate, whereas previously a platoon of workers had been needed for the job.

Despite labour-saving schemes, transport costs for charcoal wood have climbed to new heights, making up 75 % of all costs incurred before processing the raw material. In some cases these costs have caused carbonization plants to go out of business if they could not turn to other, more economical raw material sources.

Let us now look at some other factors in wood supply.

Wood attacked by fungi and mildew gives a lower yield in distillation products and charcoal. The latter, moreover, is of inferior quality, being fragile, of lower density and more readily inflammable in the atmosphere.

All the mineral substances in the wood (the chief elements being calcium, magnesium, potassium, sodium, iron, manganese, phosphoric acid) will be present in the charcoal. The proportions and composition of ash will be of great importance, especially for metallurgical charcoal.

While the basic composition is fairly constant, this is far from the case with the chief chemical constituents. Wood yields the following substances:

1. Resins, oleo-resins, gums, fatty wax, and essential oils, which are soluble in benzene, alcohol, ether, and boiling water, are known as "extractable substances". The nature of the resins and the changes they undergo during distillation have a considerable influence on the tars;

2. Pentosanes, including xylanes, arabane, and the uronic groups;

3. True cellulose equivalent to glucose obtained by total saccharification;

4. Mannanes and galactanes, always present in resinous, and often in negligible quantities in deciduous woods;

5. Lignine.

The percentage of extractable substances varies with the species from a few tenths to 25 per cent; the pentosanes vary from 9 to 28 per cent, cellulose from 30 to 50 per cent, mannanes and galactanes from 0 to 12 per cent, and lignine from 22 to 45 per cent (5).

The composition of the wood influences the yield of acetic acid and methanol, the former being chiefly furnished by cellulose and the latter and the pentosanes by lignine.

Bark contains a great deal of ash, hence barking improves the quality of the charcoal; in any case, partial barking is necessary when wood is to be seasoned naturally.

It is very important that the raw material should be stored so that it seasons as quickly as possible. Freshly felled wood has a 40 to 50 per cent moisture content, and natural seasoning brings this down to 15 to 20 per cent. The lower the moisture content, the more effective carbonization will be. Anhydrous wood, however, gives less acetic acid.

After the fuelwood has arrived at the plant site, the burden of further preparation is taken over by the personnel there. Their duty consists of three major tasks:

- Reducing the length of the felled wood to fit the loading equipment for the dryer and retorts.

- Reducing the moisture content of the wood.

- Storing the excess wood.

Therefore, the fuelwood goes first to the sawmill. Charcoal plants usually employ two bandsaws and circular saws, and a conveyor for carrying the sized logs to the retort, either directly or via a mechanical dryer.

The sawmill of a charcoal plant usually works one shift per day. Some of the larger charcoal plants have installed continuously operating automatic saws, thus eliminating labour almost completely.

The importance of the moisture content of the raw material and the relationship between moisture content and the duration of the carbonization cycle has already been mentioned. In many plants it may be preferable to sacrifice a portion of the plant capacity in order to obtain a quicker return on the money expended for wood preparation. However, one practical point should not be overlooked, and that is that dried wood is much lighter than wet wood, an important consideration for all charcoal plants where the work is done manually. An average six steres of wood, freshly felled would weigh approximately four tons, but after drying there would be one ton less to handle. If this figure is multiplied by 10 or 15 per day (the number of steres per charge) the men loading wet wood will lift 10 to 15 tons more water than men handling dry wood.

Very large industrial carbonization plants do not take any chances. The mechanical dryer forms part of their standard equipment. The principle of the wood dryer is shown in Figure 48 (8).

The logs are carried continuously to the top of the cylindrically shaped dryer by an elevator. The combustion chamber is fired with the off-gases of the retort or pyrolysis oil. The hot gases are drawn and propelled through the wood charge (up-draught system), recirculated, and the excess is expelled from the dryer by an off-gas fan. During the drying procedure, the wood charge moves slowly down to the bottom of the cylinder where it is continuously discharged.

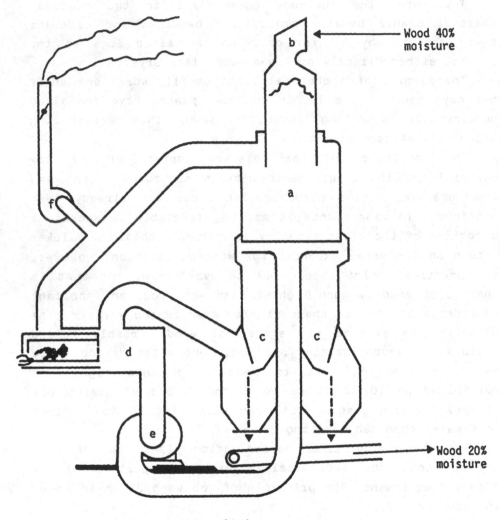

a = Drying cylinder
b = Entrance for green wood
c = Discharge locks for dry wood
d = Combustion chamber for retort gas
e = Heating gas fan
f = Off-gas fan

Figure 48. A wood dryer for continuous operation.
(Courtesy of DEGUSSA, Frankfurt/Main)

Large charcoal plants keep a reserve of wood stored nearby for periods when the wood supply stops. This can be due to climatic conditions, such as the rainy season, hurricanes, or heavy snow fall, or other causes such as transport strikes.

A retort requires a certain amount of time to cool down to normal temperature, and also needs time to heat up to its operational temperature. The hull of a retort or converter is protected by refractory lining, and this lining can be severely damaged by rapid temperature change. It follows that any breakdown in the supply of wood, with consequent cooling of the retort, causes an undesirable production standstill.

Incoming wood loads which exceed the immediate demand are normally transferred to the storage area. The storage area should be situated close to the charcoal plant so that the wood can be moved on hand-propelled trucks or carts.

The best wood storage results have been achieved with stockpilings up to a height of 8 m. The length of the piles will be limited by the dimensions of the storage area. Of course, the area should be open so as to expose the wood piles to the sunlight and wind as much as possible. If two or more wood piles have to be set up, the aisles should be kept wide open to allow the wind to pass freely. A distance of 7 m between the piles is customary.

In temperate countries, wood can be stored over 12 months in this way without danger of attack by fungi and insects. In the tropics the maximum storage time will be much shorter.

All the above applies to the carbonization practice of roundwood, firewood, branches, slabs and off-cuts of sawmills - in other words, the raw materials still in common use for most charcoal plants. Forestal residues, waste from wood processors and agricultural waste will require a different treatment, which is normally simpler and less costly.

In practice, the high cost of transporting and preparing fuelwood has made other sources more attractive for the charcoal-maker.

The cost limitations connected with the supply and the on-site preparation of fuelwood cannot be defined for all cases. A significant attitude of both large and small charcoal-makers seems to be gaining ground. This is reflected in the fact that all new charcoal enterprises are structured with a high degree of flexibility which enables them to adapt their equipment to new raw materials and resources whenever the economic situation demands a change of feedstocks.

References

(1) U.N. Conference on New and Renewable Sources of Energy, Preparatory Committee, "Report of the Technical Panel on Fuelwood and Charcoal on its Second Session". A/Conf. 100/PCI 34, 1981.

(2) L. Birritz, Industrial Management and Management Development in the Developing Countries, UNIDO report 10.476, 1981.

(3) W. Emrich, Turkey, Development of the Chemical Wood Processing Industry, UNIDO Techn. Report, Vienna, 1972.

(4) B. Mermesec, Forest Energy in Papua New Guinea, Papua New Guinea Forest Dept., Port Moresby, 1981.

(5) W. Sandermann, Naturcharze, Terpentinöl - Tallöl, Berlin, 1960

(6) D. Earl, <u>Charcoal: An Andre Mayer Fellowship Report,</u> FAO, Rome, Italy, 1974.

(7) K. I. Thomé-Kosmiensky, <u>Recycling in Developing Countries,</u> Berlin, 1982.

(8) <u>Holzverkohlung,</u> Chemische Technologie, Bd.3, Munich, F.R.G. 1971.

Chapter 6
END-USE MARKETS FOR CHARCOAL AND CHARCOAL BY-PRODUCTS

The charcoal-maker distinguishes between household markets and industrial applications; these two outlets require different tactics, have their own specifications, packing styles and marketing habits. It is, therefore, quite common for charcoal producers to be active in only one market. To facilitate comprehension, this chapter has been subdivided accordingly.

6.1 Charcoal as Household Fuel

In many countries, charcoal cookery is the only way to prepare food. Unfortunately, the price of charcoal has risen so high that in most countries only the wealthy can afford to purchase a daily supply.

In industrialized countries, charcoal cookery has moved into the public square, the patio and to the gardens, beaches and camping sites. In some countries, barbequing has become synonymous with the space age and contemporary lifestyle.

6.1.1 Lump Charcoal

Lump charcoal is still preferred by consumers since it is easy to ignite. In the market place, it is sold by volume as well as by weight.

Whereas in developing countries, the normal buyer is little concerned with the quality, size and packing style, the consumer in industrialized countries pays much attention to these aspects and distinguishes between various brand names.

Although price is by far the most important reason for selecting one or the other brand, buyers have become aware that significant differences exist in cooking times and calorific values of various lump charcoals (1).

The trend of lump charcoal consumption cannot be estimated with sufficient accuracy since statistics are rarely provided. However, it is the common belief of all persons involved in the charcoal trade and production that the demand is growing steadily.

This upward trend can be expected to accelerate in the future, owing to the fact that substitutes derived from fuel oil and other fuels have increased in price dramatically and have become unaffordable for millions of people.

6.1.2 Charcoal Briquettes

With the rising concern about energy consumption, buyers in industrialized countries are becoming aware that lump charcoal burns away fast. They are, therefore, turning their attention to the charcoal briquette which was first introduced in 1955 in the U.S.A. (1).

Since charcoal briquettes are composed of various components - binder, filler, or energy extender - they can be made long burning, hence displaying extended cooking times, e.g., twice as long as the same net weight of lump charcoal.

Unfortunately, this fuel-saving effect has remained unobserved in many developing countries until now but has caused an unsurpassed surge of the briquette market in all industrialized countries.

Other features of the charcoal briquette are its cleanliness and easy handling. Briquettes are distributed in various forms: oblong, egg-shaped, hexagonal and pillow-shaped. The latter is by far the most usable form with a maximum edge length of 50 x 60 mm and a thickness of 25 mm.

Export specifications (2)

	Lump charcoal	Briquettes Without energy extender	Briquettes With energy extender
Ashes	3 - 4 %	Max. 8 %	Max. 25 %
Moisture	Less than 5 %	5 %	5 %
Carbon	80 - 82 %	70 - 75 %	60 - 65 %
Volatiles	10 - 15 %	10 - 15 %	10 - 15 %
Binder	-	Max. 8 %	Max. 8 %
Calorific Value	6,800-7,200kcal/kg	6,000 kcal/kg	5,200 kcal/kg

6.2 Charcoal as Fuel for Industry

Before charcoal became a major consumer product for household cooking in developed countries, it was solely an industrial commodity, especially in the years when metallurgy started to emerge.

With the diversification of the chemical industry and with the increasing legislation on control of the environment, the applications of charcoal for the industrial market have multiplied.

Whereas in the household fuel market charcoal does not face competition, in almost all other applications charcoal could be replaced by natural coal, petroleum coke or lignite. The advantages of charcoal depend on six significant properties which account for its continued use in industry:

- low sulphur content

- high carbon to ash ratio

- relatively few and unreactive inorganic impurities

- specific pore structure with large surface area

- good reduction ability

- almost smokeless.

Up to 1960, charcoal was widely used for the production of carbon disulphide and sodium cyanide by the chemical industry. Although these markets have declined, they were sufficiently offset by the enlarging demand for reducing (metallurgy) and absorbent agents (filter industry).

The applications of charcoal in the various industries may be summarized as follows:

- <u>Chemical industry:</u> Manufacturing of carbon disulphide, sodium cyanide, carbides.

- <u>Iron and steel industry, metallurgy:</u> Blast iron furnaces, ferro-silica, metal hardening, non-ferrous metal industry.

- <u>Cement industry</u>

- <u>Activated carbon and filter industry:</u> Water purification, dechlorination, gas purification, solvent recovery, waste-water treatment, cigarette filters.

- <u>Gas generator:</u> Producer gas for cars, electric power generation.

- <u>Miscellaneous other applications</u>

The chemical industry and the activated carbon manufacturer prefer lump charcoal. This is partly due to their process requirements. Fine charcoal particles are more reactive but losses by emission make fines an undesirable raw material. Therefore, charcoal fines and powder are restricted to processes where compaction or agglomeration is necessary before they are brought in to the operation.

Specifications (2)

Each application has its own rigid charcoal specification which must be examined to determine both the actual process requirements and the future prescriptions due to possible alterations in the process technology. The latter has frequently had effects on the charcoal industry in the past and has led to changes in production techniques or to the shut-down of plants if not strictly observed. Charcoal quotations usually contain the general data for fixed carbon (%), ashes (%), volatiles (%), density, bulk density, moisture (%) and the sieve analysis.

It is not possible to indicate all the specifications and variations. These can be found by a careful study of the market, which is always necessary before engaging in the industrial charcoal business.

6.3 Charcoal in Metal Extraction

When iron was first made by man, charcoal was used universally as a reductant. Metallurgical coke was introduced as an alternative during the eighteenth century. Small iron blast furnaces and other reducing furnaces were at the beginning of this flourishing world-wide industry.

The charcoal-based iron industry still exists in a number of countries and continues its expansion and modernization. The centre, however, has moved away from industrialized countries. Today, Brazil's charcoal-based pig-iron industry is considered to be the greatest in the world.

Charcoal has strong reducing properties. When heated with ores containing metal, oxides, and sulphides, the carbon combines readily with oxygen and sulphur, thus facilitating metal extraction. Charcoal can therefore be used for the reduction of copper-containing ores also.

Most of the charcoal used in blast furnaces is made from hardwood species (eucalyptus). Although charcoal is generally acknowledged to be as good as, if not better than coke, there are practical difficulties in obtaining adequate supplies of charcoal to feed the large iron and steel capacitites which are needed to achieve a competitive steel price.

It is only in countries with extensive forests that the use of charcoal for iron-smelting is likely to be profitable.

6.4 Activated Charcoal

The use of charcoal for producing activated carbon is not very old when compared with its utilization in metallurgy or chemistry. Markets first started to develop around the beginning of this century in Europe.

The term activated carbon refers to various forms of carbons which have undergone a more or less intricate treatment to increase their absorptive properties.

Activated carbons are available in powdered and granular (or extruded) forms and are used in liquid and gas phase absorption processes. By taking account of the basic characteristics of the activated carbons on offer, it is found that more than seventy types are currently dominating the market.

Although the increased surface area and the absorption capacity of all activated carbons are interrelated, the size of the surface area is only one of the factors needed to measure the absorptive capacity of a given carbon for a specific purpose. The distribution of the pore volume as a function of the different pore sizes is also important. Steric effects control the access of absorbate particles to the internal surfaces and therefore an absorbate molecule may be excluded in one case but may be "fit" for other grades of active carbon.

In other words, activated carbons with large total surface areas but with micropores may be effective in removing small odour-causing impurities from gases but ineffective in absorbing large colour-forming compounds from solutions. This may explain the great number of types, grades and shapes of activated carbon on the market (3).

Production capacity for activated carbon (1979)

North America	160,000 t
Western Europe	105,000 t
Eastern Europe	20,000 t
USSR	70,000 t
Japan[x]	35,000 t

Whereas charcoal was once the only raw material for activated carbon, it has been partly replaced by other carbonaceous materials: coal, lignite, petroleum, coke, peat moss. Price considerations and the limited availability of charcoal were the main reasons for turning to other sources.

Experience has shown that there are no basic differences in the quality of activated carbons made from other raw materials. Only in the field of gas/vapour applications has charcoal-based activated carbon remained of superior quality.

Activated carbon production is in any case a low-yielding process in relation to the input of raw materials, whether charcoal is the base or not. As a general rule, the factor 1.3 can be applied to obtain a realistic figure for the necessary input of raw char.

---

[x] Several producers have subsidiaries in the Philippines and do not manufacture activated carbon in Japan.

6.4.1 Synopsis of Industrial Active Carbon Markets

Liquid phase applications

Drinking water purification, municipal wastewater and industrial wastewater treatment plants, swimming pools, tropical centres, fish tanks.

Purification of fats, oils, beverages, water purification in breweries, cleaning of bottles and winery tanks, tank cleaning in insecticide and pesticide spraying operations, cleaning of electroplating baths, dry cleaning. Decolorizing of cane and beet-sugar solutions, vitamin solutions and pharmaceuticals, high fructose corn syrup.

Gas/Vapour phase applications

Purification of exhaust emissions and immissions. Recirculated air purification. Solvent recovery in printing machinery and in processes where high volatile matter is continuously being released. Prevention of offensive odours.

Gas masks for military and civil purposes.

Other applications

Cigarette filters, catalysts for chemical processes, support for platinum and palladium catalysts, additives for food, depolarizator in air-oxygen electric cells. Additives for rubber tires, automotive evaporation control systems.

The activated carbon market has been stimulated by the legislation on water and air-pollution control in industrialized countries. Since 1977, newly introduced legislation has led to a sizeable growth of the market as a whole and is expected to increase the growth rate in the future.

Specifications

The usability of wood charcoal depends on its low ash content and its availability in uniform and unvarying quality. Exceptionally good activated carbons can be produced with charcoal made from coconut shells (4), hardwood species, sawdust and wood waste, except bark.

In every case, the activated carbon producer sets his individual standard and this is determined by his production process. The requirements however will not vary much whether the finished activated products are made for liquid, gas/vapour phase or other applications.

Although the charcoal offered to the activated carbon producer may meet all the stated criteria, the purchaser is not likely to take it without testing of the behaviour of the material in a pilot plant. Such tests comprise sophisticated procedures to determine the absorption capacity of the finished product on models. The industry has developed characteristics such as the molasses figure, the methylene-blue value and isotherms of benzene, etc. However, the charcoal supplier needs not to involve himself in these tests; in fact, he usually lacks the facilities and competent staff.

Average requirements for charcoal usable for activated carbon production (2)

Fixed carbon	82 % min.
Ashes	4 % max.
Volatiles	10 %
Moisture	4 %
pH	4 - 10

pH refers to an acidity test of a water extract from the charcoal.

6.5 Speciality Markets for Charcoal

Horticulture

Charcoal is used in different grades as a top dressing for the improvement of lawns and bowling greens. These top dressings act as mulch and also provide valuable trace elements as well as soil sweetener.

Pigments for printing and paints

Vegetable blacks are dense and show great strength.

Poultry and animal feed

Feeds are sometimes supplemented with charcoal fines to prevent certain diseases.

The above speciality markets are limited. They can, however, provide a good support to the charcoal-maker in developing countries also, if properly used.

6.6 Charcoal for Producer Gas

Because of the world energy crisis, another application is attracting much interest, namely the generation of gas from charcoal. Producer gas is best utilized as a fuel for cars, trucks, and farm vehicles in particular, but is also suitable for gas engines used for the generation of electricity.

Producer gas made from charcoal is not new, it was in common use during the last war in some European countries where it filled the gap in petrol supplies. At that time, wood was also in widespread use as a fuel for gas-propelled car engines (5).

When compared with wood, charcoal displays several advantages which make it a superior fuel:

- high attrition resistance against pressure

- high energy density of 7,000 kcal/kg as compared with 4,200 kcal/kg for dry wood

- low moisture content of 4-5 % as compared with 20 % for air-dry wood

- no predrying necessary because of the low water content

- the gasifier and storage container can be kept small, as compared with wood generators.

At present, it is not possible to forecast the market for producer gas charcoal. Figures for the demand have been published sporadically and rarely. To the knowledge of the author, the interest in developing countries is considerable. Progressive developments have taken place already in South America and in East African countries. All projects are aiming to make farm vehicles independent of expensive petrol supplies. Other goals are to generate electricity for sawmills or villages in rural areas.

It is recommended that the charcoal project planner should devote part of his attention to the use of charcoal as producer gas.

In practice, the number of equivalent cubic metres of producer gas which can be obtained is:

1 kg wet wood = ca. 1.25 - 1.5 m^3 producer gas
1 kg dry wood = ca. 1.9 - 2.2 m^3 producer gas
1 kg charcoal = ca. 4.25 - 4.75 m^3 producer gas.

6.7 By-Product Utilization

Pyrolysis Oil as a Fuel

It has been commercially demonstrated that pyrolysis oil is usable as a liquid fuel and also has a potential as a raw material for the production of acetic acid, methanol, tar oils, creosotes, tar, etc. (6).

The oil has been used commercially as a fuel in cement kilns, steam boilers, brick and lime kilns, etc. Its potential as a fuel resides in the fact that the heating value can be significant (between 3,000 and 4,000 kcal/kg); other advantages are that the viscosity can be controlled by adding water (up to 25 %) and it is transportable in drums and tank cars. The unlimited miscibility with fuel oil No. 6 permits its use in blends.

The applications of pyrolysis oil are manifold and can be found anywhere, especially in industry. It is commonly believed in the charcoal industry that the utilization of pyrolsis oils will grow in the future as the facts and opportunities become public knowledge. The economics of existing plants with recovery of pyrolytic oil have demonstrated the superiority of these processes and proved them to be an excellent investment for the owner.

Experience gained in commercial operations has brought out some important findings:

- The oil should contain a minimum of 15 % water and a maximum of 30 %. This serves to reduce viscosity and promotes better atomization. Also, it appears to enhance combustion.

- The oil should be heated to 75° C to enhance flammability.

- Precautions must be taken in the heating system so that the oil cannot be advertently heated above 110° C.

Pyrolysis oils are corrosive to mild steel, but the corrosion rate is a fraction of a mil per year for copper and 304 stainless steel. Therefore, before using it in an existing regular burner, all parts exposed to the oil must be replaced by corrosion-resistant devices.

Other characteristics are:

- The oil is heavier than water and forms a stable emulsion with water.

- The viscosity of the oil at a moisture concentration of 10 per cent or above is lower than the viscosity of No. 6 fuel oil.

- The flash point is high.

Pyrolysis oil as a feedstock for industry

Economic activities in silvichemicals showed divergent trends until 1970. References were made solely to the use of charcoal activated carbon, naval stores, rosin, tall-oil fatty acid products pyrolysis oil was neglected.

Pyrolytic silvichemicals experienced a peak period before and during the First World War. They provided the major part of the raw material for the rising chemical industry. After that, synthetic products made from mineral coal, crude oil and natural gas became the front runners and were produced at low cost in very large quantities. Only during the last ten years have the chemical and food industries started to turn their attention back to pyrolytic silvichemicals.

Although the fractionation and extraction of pyrolysis oil are not highly sophisticated operations, they do require skilled labour. The normal approach of the charcoal industry to this technology is first to produce crude pyrolysis oil. The investment for this expansion is not high and it can be done gradually as the markets demand it.

The number and yields of silvichemicals obtainable vary widely according to the composition of the raw materials. But several of them can be extracted from almost any pyrolysis oil and they are, therefore, described briefly under here.

- Acetic acid: is recovered as raw acid, a yellow to dark-brown liquid. Several commercial grades can be made from it - technical acetic acid, wood vinegar for the food industry, glacial acetic acid which solidifies at 16.6° C. Specifications for grading vary from country to country.

- Methanol: can be separated by rectification from the raw acetic acid; it is dark-yellow in colour and contains small quantities of acetic, butyric and propionic acids. Commercial grades are pure methanol (an alternate fuel for cars and commercial vehicles), wood spirit for denaturation, chemical solvent, and basic products for various chemical processes (7).

- Pyrolytic tar: can be obtained by settling down from concentrated pyrolysis oil and is derived at different stages of the refining processes for raw acetic acid and raw methanol. The commercial grades of pyrolytic tars are distinguished by the flash points (60 - 133° C), viscosity and calorific values (5,800 - 7,600 kcal/kg). They are also differentiated by their solubility in water.

- <u>Pyrolytic pitch:</u> a residue of tar distillation, has a dark-brown or black colour. Charcoal plants normally supply four different grades which are marked by their softening points.

- <u>Tar oils:</u> are recovered at various stages of tar distillation and during raw acetic acid refining; they range in colour from transparent to dark-brown and have a characteristic smell. More than 10 commercial grades are known and made according to specifications, classified by specific weight, viscosity, calorific value and boiling point diagram. They are widely used for the preservation of wood and in the flotation process for the separation and beneficiation of ores.

- <u>Creosote:</u> the pyrolysis creosote is a mixture of guaiacol, creosol, phenols, and phenol ethers; a clear, light-yellow liquid, it must not turn dark if exposed to direct sunlight. This is a high-priced product, and specifications are governed by national pharmacopoeia.

<u>Uses for retort or converter gas</u>

Applications for the residual gases of a charcoal plant have a history dating back to the middle of the nineteenth century. At that time, several cities in Austria made use of them for public lighting.

In common usage, we only consider the off-gas which leaves the plant exhaust pipe after the vapours, which make up the composition of the pyrolysis oil, have been extracted from it. In other words, the retort or converter gas consists of the uncondensable parts.

Measurements and observations made in commercial operations over the years suggest that the average composition of the residual gas (using wood as raw material) is: (6)

CO_2	59.0 Vol. %
CO	33.0 Vol. %
Methane	3.5 Vol. %
Hydrogen	3.0 Vol. %
Vapours, etc.	1.5 Vol. %

However, these figures will deviate significantly from the gas produced in continuous charcoal operations. In a batchwise operated retort, the initial gas consists of air and water, followed by a gas with high CO_2 and CO contents and during the last third of the carbonization cycle, the gas is loaded with hydrocarbons and vapours and has a low flash point. In contrast, the off-gas expelled from continuous operations is uniform and its content of combustible substances depends strictly on the gas temperature and raw material congruity.

From what is said above, it can be deduced that there cannot be a single calorific value of retort or converter gas. In fact, values range between 850 kcal and 1,600 kcal per m^3 (at 15° C).

In the first place, the gas is used in the carbonization plant to heat the raw material dryer or as fuel for the charcoal briquette dryer. Another very popular use is to burn it in an auxiliary steam boiler. The generated steam is sold to another industry.

6.8 Synopsis of Major Uses of Charcoal and By-Products (1)

Product	Raw material	Application
Charcoal, lump	Hardwood, softwood	Activated carbon, ferro-silicon, grill coal, metal working, sodium cyanide, carbon disulphide, Swedish steel, silicon

- 193 -

Product	Raw material	Application
Charcoal granular	Charcoal, lump	Activated carbon, additive to animal food, fillings compound for bottled gas, hardener
Charcoal, dust	Charcoal, lump	Activated carbon, lining of moulds in metal foundries, production of briquettes, cementation granulate, pyrotechnics
Pyrolytic oils	Hardwood, softwood agricultural wastes	Fuel for steam boiler furnaces, solid fuel, fired engines, metallurgy, fire-brick factories, etc., raw material for chemical industry
Wood gas	Hardwood, softwood agricultural wastes	Heating gas for all types of operations using solid or liquid fuels, gas engines
Wood vinegar	Hardwood, softwood	Preservation and flavouring of meat and smoked fish, perfume and aroma industry
Wood tar	Hardwood	Rope industry, veterinary medicine, pitch, creosote
Crude methanol	Wood alcohol	Methyl acetate

Product	Raw material	Application
Solvent	Wood alcohol	Cellulose esters and agglutinants, synthetics, lacquers
Methyl formate	Crude wood vinegar and crude methanol	Cellulose esters and agglutinants, synthetics, lacquers
Methyl acetate	Crude wood vingegar and crude methanol	Cellulose esters and agglutinants, synthetics, lacquers
Acetic acid	Crude acetic acid	Chemical, pharmaceutical, food, convenience food, rayon, textile and film industries, vinegar
Propionic acid	Crude acetic acid	Pharmaceuticals, flavour and fragrances
Butyric acid	Crude acetic acid	Pharmaceutical and perfume industries

6.9 Charcoal Costs and Fuel Prices

The prospective charcoal-maker often wishes to invoke approval of his project by a comparison of his expected charcoal costs with the prices of other fuels on the market.

Charcoal projects are normally based on one or more of the following concepts:

- Production of charcoal as the sole fuel.

- Production of charcoal associated with by-product recovery.

- Production of charcoal as a household fuel for use in developing countries.

- Production of charcoal to make a profit in the existing domestic market.

- Production of charcoal to make a profit or to earn foreign currency in export markets.

It is obvious that each of these concepts, by itself or in combination, will require alternatives in the selection of feedstocks and technology. Because universal charcoal costs do not exist, it is necessary to carry out a thorough feasibility study of individual projects to protect the investor from unpleasant surprises.

Comparison with other fuels - firewood, kiln charcoal, kerosene, natural gas, bituminous coal, etc. - is not an easy task and can only be accomplished by considering the energy values of the compared fuels.

To obtain reliable values of fuels for household use, a test under confined conditions will best answer the question. Especially the efficiency of cooking fuels will be affected by the devices in which they burnt - shape, size, draught, etc.

In order to make definitive statements concerning the position of charcoal within the class of combustible fuels, a comparative evaluation must be made. It is here that reclassifying biomass with fossil energy for major uses becomes important and that the analysis of fuels becomes critical.

The following table presents four fuel families from dried sewage sludge to bituminous coal:

Fuel Material	Ashes %	Moisture %	Higher heating value	
			KJ/kg	kcal/kg
Premium fuels				
bituminous coal	4.1	2	31,300	7,500
bituminous coke	1	-	26,700	6,400
charcoal, lump	3	5	30,000	7,200
charcoal, briquettes	9	4	27,000	6,400
Wood fuels				
hardwood, green	1.5	37	10,500	2,500
hardwood, dry	0.5	15	15,500	3,700
pine bark	2.9	3	20,000	4,800
sawdust, fresh	1.8	45	9,600	2,300
sawdust, dry	0.8	14	14,600	3,500
Waste-based fuels				
coconut shells	2.5	12	14,650	3,500
bovine manure	17.8	6	15,900	3,800
bagasse, wet	1.3	45	8,350	2,000
bagasse, dry	1.9	11	15,900	3,800
coffee husks	1	65	6,550	1,550
rice hulls	15.5	6	15,000	3,600
rice straw	19.2	6	14,800	3,550
paper-mill sludge	10.2	10	12,100	2,900
sewage sludge	17.4	12	4,700	1,120
Fuels with high hydrogen content				
butane/propane	-	-	45,600	10,900
kerosene	-	-	44,300	10,600
diesel oil	-	-	40,100	9,600

On the basis of this fuel table, certain statements can be made concerning the relationship of charcoal to the total family of combustible fuels. The primary fuels are those in which both carbon and hydrogen contribute significantly to the energy value. The wood fuels can be considered as transitional fuels between the categories of "premium" and "waste-based fuels".

For the development of a charcoal cost projection, it is important to establish whether charcoal displays a closer proximity to the premium or transition fuel category.

To fix the position of charcoal within the premium fuel group, it is necessary to look at the potential pollution analysis. Traditionally, sulphur and ash are considered the principal impurities in combustion fuels. The sulphur, on combustion, forms SO_2 which is a pollutant. It can also combine with rain to form dilute sulphurous acid, or can be transformed into other potentially dangerous compounds.

When these pollution criteria are applied, the value of charcoal can be appreciated, and it should thus be classified under the top grade fuels shown in the table, if regarded solely as an industrial fuel.

The present price levels for 1 million kcal bear out this statement:

Charcoal:	U.S.A.	U.S.$	31.40
	Europe	U.S.$	45.60
Heating oil:	Europe	U.S.$	34.70

The attractive price levels which all charcoal products are enjoying in industrialized countries do not permit their application in processes where the nature of the product can support the cost of expensive fuels - e.g., the extraction of rare metals from their ores, or for hardening quality steels in metallurgy.

The picture changes when one turns to charcoal-making in developing countries. If we assume production costs of $ 55 - $ 100 per ton, the purchase price of 1 million kcal of charcoal energy will be between $ 7.80 and $ 14.30. It is obvious that charcoal for domestic use is competitive with almost any other fuel of the primary and premium groups, although fuel prices may vary greatly from country to country. However, it must be pointed out that when comparing charcoal prices with fuels of the premium group, the fixed carbon content of charcoal must be taken into consideration, and this normally fluctuates between 68 % and 85 %. The necessary correction is given by the following formula:

Charcoal: US. $ 55/ton   fixed C = 70 %
Charcoal for
metallurgy: US. $ ......   fixed C = 85 % (spec.)

$$\frac{\$ \ 55 \ x \ 85}{70} = U.S. \ \$ \ 66.79 \ per \ ton$$

Based on fuel costs, the use of industrial charcoal can be recommended for iron, nonferrous metals and steelmaking:

- extraction of iron from ores in blast furnaces and pig-iron electric furnaces

- extraction of manganese and ferro-silica from ores

- pig-iron for steelmaking and for foundry iron

- sintering processes

- steel hardening.

Charcoal has other potential uses in developing countries engaged in the build-up of their own chemical industries. Here it is important to note that the chemical reactivity of charcoal has been commercially proved. In such applications, the use of charcoal will be judged not by its fuel value alone, but also by physical and chemical properties, such as porosity, density, congruity of the ash content, hydrophilic and hydrophobic characteristics.

6.10 Packing and Shipment for Export/Market Strategy

The packing style is of great importance to attract the buyer and to protect the merchandise.

In recent years, 3 kg or 10 kg paper bags have become popular in industrialized countries, whereas the jute bag is common in developing countries. Paper bags have a rectangular shape and a flat bottom which allows the bag to be set in an upright position. Bags are made from two-layer or three-layer paper. The outside is normally three-colour printed and carries brand names, advertisements, or special warning labels wherever this is required by law.

For shipment; two or more bags are packed together in one bale. The baler bag consists of strong craft paper which is usually left unprinted and in natural colour. For identification purposes, the baler carries the denomination of the contents on a slip affixed to the outside.

Baler packing is used whenever the charcoal is "dead stacked" for transportation. The most frequently requested means of transportation is "on pallets". Since two or more pallets are placed on top of each other, care has to be taken to prevent the charcoal bags from damage as a result of unavoidable shifting of the pallets during transportation; this is one of the most frequent causes for damage claims. The best known method at present is a box carton which is placed over the pallet load, thus protecting the content.

Overseas shipments are made by containerization only. The 40-foot Dry Cargo type, an enclosed container with one large opening at the front side, is the preferred size.

Approximately 70 per cent of all charcoal for the industrial market is transportet in bulk quantities by truck, trailer or rail. These shipments need good weather protection (2).

For the transportation of charcoal powder, special silo or tank trucks are available. Charcoal fines and dust are also frequently shipped as open deck loads. Extreme care must be taken to keep the entire load wet, primarily to prevent self-ignition but also to avoid dust nuisance.

Expenses for damage, demurrage charges caused by inappropriate or delayed deliveries are customarily reimbursed by the charcoal supplier and fall under his responsibility.

Freshly produced charcoal is not ready for immediate bulk shipment because of its tendency to absorb oxygen which frequently causes fires. A reasonable curing time is therefore necessary. Forwarders do not usually accept shipments without certification of sufficient curing time. Charcoal briquettes, lump charcoal, charcoal fines and powder are not classified as self-igniting goods, but are listed in this category. That means they may be shipped without special transportation permits, but must not be carried in contact with flammable goods such as wood products, etc.

Marketing and business strategy

When one considers the variety and particular exigencies of the markets, it becomes evident that the newcomer must surmount many difficulties to find his place. Even the charcoal business can be commonly considered as a "buyer's market".

There are many differentiations between the various applications and the constant requirement for reliable supply and uniform quality.

Another task is to follow-up the fast-changing trends of technical change in processes; this requires frequent alterations of products which can have a strong impact on costs.

The legislative amendments for trade, transportation and product liabilities are another concern.

Customer contact and structuring of the annual market prices are very sensitive matters and these should be dealt with on a strictly personal basis, in an atmosphere of mutual confidence.

Finally, the new entrant needs to become familiar with the attitude and behaviour of his clients.

All these facts have led in time to the pursuance of proved market patterns to find optimal ways for product promotion. Therefore, the charcoal supplier who cannot fall back upon his own expertise or sales force will normally engage the services of a representative or an agency who can provide the business and technical know-how. These services include:

- Market survey.

- Material tests.

- Advertising.

- Client contacts.

- Price structuring.

- Preparation of sales contracts,
  Applications for customs, import licences, etc.

- Obtaining sales certificates stating that the products meet health and safety standards and legislative requirements.

- Handling customer complaints

- Obtaining bank approvals for customer credit-lines.

Since the agent or representative acts as a local broker, he receives a commission to reimburse him for his expenditures and for remuneration. Fees are based strictly on sales results and are set out in an annual contract. However, the eventual success of a charcoal business will depend on the flexibility of the supplier and how quickly he can serve the needs of the customers.

Last but not least, success will be related to the technical capability of his production facilities. In other words, the design of the charcoal plant may be the limiting factor.

6.11 World Production

In many countries, carbonization products represent a major factor of domestic trade. Since the raw material exists in these countries, the products may be shipped without laborious injunctions, and this makes them a stable merchandise.

The world charcoal production cannot be calculated with precision, nor are the quantities of produced and consumed charcoal by-products known, as only a few countries provide statistical data.

In developing countries, consumer research in the charcoal market is rarely undertaken. The most elaborate inquiry into the level of charcoal production in developing countries was carried out by FAO in 1980, with the object of providing the best possible estimates to the UN Conference on New and Renewable Sources of Energy in Nairobi (1981). Data were gathered through questionnaires and by searching available reports.

Although much effort went into the survey, it was found that for many countries there was a wide range of estimates and very few countries had reliable data available. Subsequently, it was decided to treat the available estimates as random observations.

Having been involved for more than two decades in pyrolytic research and production, the author has done a great deal of market research and gathered results and studies. However, these surveys are more of a punctual nature than global and they relate only to the situation in particular countries.

Despite these drawbacks, it is possible to build up a more satisfactory picture of today's charcoal production by considering also the available import figures, evidence of charcoal shipments, the experience of equipment suppliers, etc.

Estimated Annual Charcoal Production
(Basis: 1981)

Area	.000 t/y	Remarks
Africa		
East African countries	150 - 170	
Madagascar, Mauritius,		
South Africa	85	Includes charcoal briquettes.
West African countries	580 - 600	
Americas		
Argentina, Brazil	4,900	Includes charcoal briquettes.
Canada, USA	1,400	85 % charcoal briquettes
Central America	25 - 30	
Asia		
People's Republic of China	200 - 450	
Philippines, India,	100 - 140	Mainly coconut shell charcoal, including charcoal briquettes.
Sri Lanka		

Estimated Annual Charcoal Production
(Basis: 1981)

Area	.000 t/y	Remarks
Australia, New Zealand	70 - 90	
Europe		
EEC countries	130	Includes charcoal briquettes.
Czechoslovakia, Hungary	130	Includes charcoal briquettes.
Romania, Yugoslavia, Scandinavia	30	Includes imports from S. America
Poland, USSR	250 - 300	Not including Asian territories.
South Pacific Basin	10 - 12	

Although the figures in the table are partly derived from estimates, they illustrate the charcoal situation in general and show South America as the largest charcoal-producing continent.

FAO attempted a forecast in the above-mentioned study by comparing the latest production figures with those of 1970. The results show clearly that demand and production of charcoal has increased in all the countries searched. In fact, in no case has a decrease of charcoal production been found between 1976 and 1980.

Although there is no way to quantify the upward trend of charcoal output, it is obvious that the surge for pyrolytic commodities has been accelerating since 1972, when the world-wide energy crisis became visible.

In addition, the desire of people in industrialized countries for increased and improved charcoal cookery, the need for producer gas propelled engines and the recognition of the value of pyrolsis oil as a substitute for fuel oil has already given a further impetus to charcoal production.

Presently only a few developing countries are servicing with their exports the markets of industrialised countries.

But there is a growing awareness at the governments and private enterprises of the potentials for cropping hard currency by foreign charcoal trade.

On the other side the charcoal industry, charcoal traders and shipping companies of the industrialised countries have become very observant to the existent opportunities with accessible partners there.

This trend becomes visible in the growing flood of inquiries which reach the desks of all people concerned with the business.

Charcoal exports have been curbed in most cases by failing to meet the quality standards, by lack of available packing materials and unreliable supply in terms of meeting fixed shipping dates.

It is well understood that the charcoal consuming will increase in all developing countries with the improvement of living conditions because charcoal is a convenience fuel for the households and the advantages must not be repeated here.

However, the important questions spirating around the implications for the domestic markets in developing countries will remain unanswered as long as these countries are lacking of the respective statistical material and a national charcoal programme.

References

(1) H. Messman, "What is Charcoal?", paper presented at 12th Biennial Conf. of the Institute for Briquetting and Agglomeration (IBA), USA, 1971.

(2) W. Emrich, The Charcoal Markets in Industrialized Countries and the Impacts of Charcoal Exports in Developing Countries, FAO Report, Rome, 1981.

(3) J. Hassler, Purification with Activated Carbon, Chem. Publ. Co., New York, U.S.A. 1974.

(4) J. Woodroof, Coconuts: Production, Processing, Products, AVI Publ. Co., Inc., Westport, Conn., U.S.A. 1970.

(5) L. Jaeger, Grundlagen der Holzgasanlagen fuer ortsfesten und fahrbaren Betrieb, Muenchen, F. R. G., 1935

(6) M. Klar, Technologie der Holzverkohlung, Berlin, 1910.

(7) E. Plassmann, On the Trail of New Fuels, VW Research Center, Wolfsburg, F.R.G, 1974.

Chapter 7
PLANNING A CHARCOAL VENTURE AND SELECTION OF EQUIPMENT

In today's fast-changing world with its great need for renewable energy sources, it is essential that the prospective charcoal-maker be aware of the limits and opportunities of the technology and the choice of equipment available to him. An important criterion will always be the achievable energy yield from the raw materials under consideration. Yield comparisons are difficult to establish; a great deal of experience is necessary, and tests must be performed on firm parameters. These are, among others, dry material weight, fixed carbon content and calorific value of the char, and a defined energy value of the liquid and gaseous by-products, if any.

It is also well known in the charcoal industry that very often the yield figures presented are not sufficiently reliable and hence can be very misleading.

European experience has proved that, as a general rule, the energy recovered from a raw material does not exceed the following values:

Simple charcoal kilns: charcoal pits, earthmound kiln	18-22 %
Brick and metal kilns	24-28 %
Retorts without by-product recovery	30-35 %
Biomass converter with recovery of by-products	65-80 %

These data were obtained from hardwood species not common in tropical countries. Also they cannot be applied to the varieties of agricultural residues. They should serve the planner more or less as rough guidelines.

However, it is obvious that modern technology offers the significant economic advantage of superior raw material utilization as compared with traditional charring methods. The benefits of industrial charcoal making include increased profits from the fuller use of forest products as well as the saving of silvicultural cost and natural resources. Advantages and rewards will not depend on the size of the charcoal production, whether it be a small-scale plant in a remote sawmill or a large operation within an industrial compound. They will rather be related to the way the planning, the design and the organization have been executed.

National planning

In developing countries, particularly where charcoal is to become a basic industrial raw material on a large scale, the planning should ensure that long-term wishes are likely to be satisfied.

It is certainly not necessary for the small entrepreneur, including the itinerant charcoal-maker, to worry about long-term planning, but it is essential that the Planning, Forest or Agricultural Department which supplies the raw materials ensures that the proposed plant will work for the benefit of the country. The work of planning a charcoal industry should be carried out at two levels:

1. A national plan for the charcoal industry which must fit in with the development aims of the country.

2. Planning for particular projects which must conform to the objectives of the planning principles laid down above. Planning at the national level will take into account the annual increment of raw materials which can safely be converted into charcoal, pyrolysis oil and industrial heating gas. It also focuses on the probable production cost, transport and marketing organizations required. This full appraisal should therefore be done

by an economist, working either directly out of the National Planning Department or as a liaison, coordinating the interests of various departments.

Commercially proved techniques are available to carbonize even small quantities of feedstocks economically. This fact will certainly become of much importance for the planner and entrepreneur in developing countries with agricultural wastes in abundance.

Another trend, already established in industrialized countries, is forecast to spread widely: this is the use of thinners (energy extender) by charcoal briquetters. Initially, thinners were intended to stretch the precious raw material reserves, in other words to gain more benefits from less carbon. But markets have adapted to the "long cooking" fuels and customers know how to use them in favour of the family budget. We are confident that this trend will catch on in developing countries also, as soon as governments become aware that their natural resources can be extended to serve a broader population or to raise exports.

The greatest impact may be expected from the development of "Integrated Charcoal-Making Concepts". In a broad sense, this term applies to projects attached to an existing industrial system or to one still on the planner's desk.

Whereas until the end of the Second World War, a charcoal operation was based solely on the availability of wood, future charcoal plants will be set up primarily to serve a predetermined purpose. In other words, the future plant will be designed to make the best use of the raw material reserve by supplying a variety of products. The keyword for future plants will be their interlocking capability with other industries and energy consumers, and fitting into a state or country-wide household supply programme.

7.1 Planning of Projects

A modern charcoal industry can be designed for integration in existing or future projects, e.g., a charcoal iron industry. It is necessary to carry out a project appraisal in order to ascertain that it is economically and socially sound. A commonly adopted sequence of stages in a project appraisal is the following:

- Study of project objectives and alternatives.

- Appraisal of suggested technology and its proof.

- Social cost/benefit analysis.

- Commerical and financial analysis (including sensitivity and risk analysis).

- Other economic considerations.

Conclusions and recommendations

The above-mentioned Integrated Charcoal Concept depends on the local situation and therefore Figure 49 can only convey ideas.

A charcoal plant may deliver its char to an activated carbon producer, an iron or steel mill, or make briquettes for the domestic and export markets, etc. At the same time, it can supply pyrolysis oil as a fuel for a brick factory, particle board plant, distilleries or/and industrial steam boilers. The options are numerous and cannot be described in detail in this handbook.

If the decision is made that the development of a charcoal industry or its expansion is likely to bring socio-economic benefits to the country, a government department, most likely the planning department, should

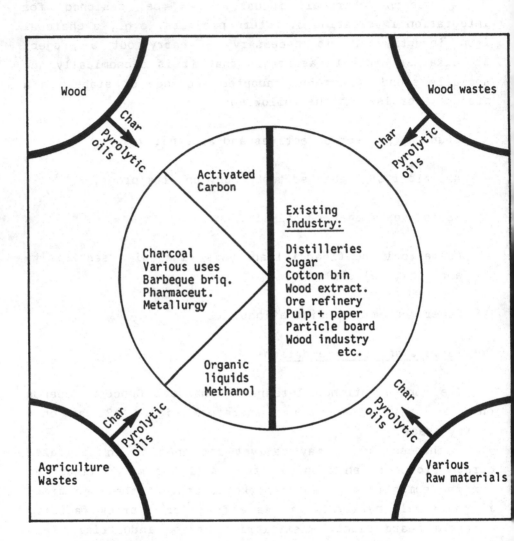

Figure 49. The integrated carbonization concept with four carbonizers.

appoint a coordinator with interests in forestry, agriculture and chemistry to head the new charcoal section.

This section would be given the task of ensuring that maximum effort is put into the implementation of the decision to institute or expand the charcoal industry. The coordinator, whose suggested title would be charcoal development officer, would have the following duties:

- Investigation of all possible raw material sources.

- Survey of the local and export markets.

- Designing a research programme.

- Selection and hiring of consulting services.

The charcoal development officer would also be concerned with the training of local staff. A training programme will comprise:

- Basic explanation of the principle of carbonization.

- The correlation between raw material type, preparing and drying of feedstocks and its effect upon the yield of char, pyrolysis oil and gaseous energy.

- Practical demonstrations on a pilot plant (usually not available in developing countries) or the construction of such pilot projects may be the major objective of the course.

- Safety precautions and first aid.

- Costs, bookkeeping, and marketing.

Each course should be terminated by an examination at

which the standard of marks for awards should be set fairly high.

The decision to set up a properly organized charcoal industry should be made before the organization of a charcoal research and development team. Information gained from research and development will lead to modification of some of the suggested projects and therefore close consultation and liaison with the Planning Department is essential.

At present chemical charcoal research is concentrating on the identification of compounds in the pyrolytic oils from all materials. Research had been dormant during the long period of cheap petroleum. Today interest is focusing on the ingredients of pyrolytic oil as rare raw materials for the chemical, pharmaceutical and cosmetic industries to substitute for compounds which have their origin in the synthesis of petroleum derivates. Some objectives, however, are farther reaching. Extracts of pyrolysis oils can serve as an alternative fuel for cars, trucks, agricultural equipment and electric generators, if properly prepared and gasified.

Cogeneration of energy has again become an essential part of the charcoal industry. Numerous combinations are possible and can be developed. The credits gained from the sale of by-products can set-off a major part of operational costs and it appears that these contributions will become more attractive in the future.

The case studies given in Appendix 1 (CISR-Lambiotte plant and Vertical Flow Converter (Tech-Air) plant) bring out the main characteristics of a modern plant and give an insight into the planning process.

7.2 Selection of Charcoal Equipment

The heart of the charcoal plant is the kiln, retort or converter. In a strict sense, there are no perceivable variations between retorts and converters.

In any event, the yield of charcoal, the uniformity of its quality, the quantity of liquid and gaseous by-products, the throughput and output of the plant will be determined by the conversion equipment selected. If a decision has been made about the type of operation, the other parts of the factory must be adjusted to it.

On the other hand, the raw material type - long or short wood logs, fine or coarse material, agricultural or non-agricultural, will influence the choice of kiln, retort/converter and the kind of operation - batchwise or continuous.

It was discovered very early that the energy demand of a charcoaling process is closely related to the retort capacity, if the operation takes place batchwise. However, the yields of charcoal and by-products are very little affected by alterations of the retort capacity. The same phenomenon can be observed with circular, well insulated earthmound kilns, although reliable measurements are extremely difficult. The table given in the introduction to this chapter summarizes data which were collected from European plants operating with hardwood species.

The energy demand of the charcoaling process is related not only to retort capacity and moisture content of the raw feed, but also depends on the efficiency of the heat exchange between the ambient heating gas and the surface of the raw material parts or particles. In a batchwise operated retort or converter, the individual parts of the raw material remain almost immobile, or are at best in very slow motion during the entire carbonization cycle of many hours or days. Under these conditions, access to the surface of the raw material parts will be made more difficult. Some of the parts will also remain close together and will not be sufficiently exposed to the heating gases. This effect is frequently observed in poorly operated charcoal kilns and results in the discharge of a high proportion of unburned or half-burned wood logs.

Heat exchange efficiency is also the reason why most of the retorts and almost all kilns require, for their proper charging, specific dimensions for the diameter and length of the wood logs. Otherwise, penetration of the heating gas will not be adequate to meet the necessary velocity rate of heat exchange because of the high density of the charge.

Considerable improvement of the heat exchange can be achieved if the feed is kept in steady movement, e.g., in a continuously operated converter plant. This energy-saving effect is demonstrated in the following table for the continuously operated SIFIC retort in comparison with the batchwise operation of a Reichert retort. In both cases, log sizes and moisture content of the feed were identical (1).

Yield of:	SIFIC retort kg/t dry wood	REICHERT retort kg/t dry wood
Charcoal	330 - 350	330 - 350
Methanol	23	19 - 23
Acetic acid	77	66 - 70
Heating demand	17 kcal/kg dry wood	340 kcal/kg dry wood
Electri. demand	5.6 KWh/t dry wood	21 KWh/t dry wood

Insulation of the kiln, retort or converter becomes an important issue in all countries with seasonal temperatures below 15° C or in locations where high wind-chill factors are prevalent. Losses of tangible heat are usually negligible in hot climates, even at night.

There is a major difference between systems which heat the wood by external means, using wood, oil, gas, etc., and systems which allow combustion on a limited scale to occur inside the carbonizer by burning part of the wood charge and using this heat to dry and carbonize the remainder. The latter method should be the most efficient since the heat is generated where it is needed, using low-cost wood fuel. In practice, it is difficult to control the combustion and some extra wood is burned which lowers the yield.

The uniformity of the quality of the produced charcoal and by-products is an important consideration for the consumer. As one regards the fixed carbon content of the charcoal, this is directly related to the temperature course, the terminal temperature of batchwise operations and the residence time of the raw material. Therefore, the well designed charcoal retort or converter can be judged by its temperature control installation, the flexibility and sensitivity to alterations.

Kilns as a means of traditional carbonization are internally heated and manually controlled. The three possibilities found are earth, which is lowest in cost, bricks or masonry of intermediate cost, and steel which is the most expensive. Steel kilns are further subdivided into portable and fixed types.

Portable steel kilns possess two advantages: they can be moved easily, which may be useful, and they cool quickly, allowing a shorter cycle time. However, portability is not always beneficial, since it becomes difficult to organize and supervise production efficiently; moreover, fixed brick kilns can be cooled quite rapidly by injecting water into the kiln (with care!). Although cycle times are still around six to eight days, compared with two for steel kilns, the greater volume and much lower cost of brick kilns make them preferable except where portability is essential.

Earth kilns and pits, even when operated efficiently, are slow burning and slow cooling and contaminate the charcoal with earth. However, where capital is limited or non-existent, they have real advantages.

It has been mentioned before that the type of raw material and the size of individual parts or particles may demand a certain kiln, retort or converter. As a rule of thumb, raw material sizes not exceeding 100 mm in length and with a maximum diameter of 25 mm cannot be satisfactorily converted in a batchwise operation. Such raw materials may have been discharged by sawmills in the form of chips or sawdust, or by plantations and agricultural crop processors in the form of nutshells, husks, hulls, leaves, twiglets, pulpa from coffee or bagasse from the sugar industry. They may also have been produced artificially by wood chippers, farm shredders, etc. In any event, they tend to cause high density charges (in some cases they have to be compacted first) when fed into the converter, consequently reducing the heat exchange efficiency significantly.

The equipment supplier is therefore obliged to design and manufacture special machinery for the charcoal producer.

Basic equipment list

Raw material receipt and preparation: chain saw, circular saw, roller band saw, chipper, shredder, hammer mill, drag chain, screen. The feasibility in each special situation must be proved by a test.

Raw material dryer: band dryer, screw dryer, single and multi-pass rotary dryer, dryer of a special type (Lambiotte), grain dryer, pneumatic dryer, conveyor belt, elevator, electro-magnet for the removal of iron debris, silo. Drying tests are strongly recommended where sun or air drying is not sufficient. When the moisture content of the raw feed is less than 20 %, drying is not usually necessary.

Off-gas system, pyrolysis oil recovery: scrubber, water cooler, demister, draught fan, valves, temperature control, oil pump, piping, tubing, storage tank. Major parts of the off-gas system must be made from copper, stainless steel, ceramics or wood. The size of the scrubber and the surface area of the water cooler must be determined individually.

Flare stack: chimney, automatic burner, burner chamber, pick-up for heating gas supply, draught fan. If burning of excess gas is required by environmental legislation, a special design is necessary.

Not all the items in the above list will be necessary in particular plants. Since an industrial charcoal plant can use machinery from other industries with minor alterations, it has become common practice to employ second hand equipment as much as possible; this could reduce the initial investment costs considerably.

Traditionally, the inherited wisdom of rural societies has played an important part in charcoal-making. To use the established method which is known to work successfully in a locality is the logical option for those who cannot afford to take risks because of their precarious economic situation. Where social factors are dominant, it is usually very difficult to introduce a new technology of charcoal-making unless the social factors are changed. Frequently one sees attempts to modify the technology of charcoal-making by providing aid: inputs such as chain saws, new kilns and so on. When these inputs are no longer available, economic necessity forces the producers to revert to the traditional, successful method with all its obvious technical faults. Therefore, carbonizing methods cannot be evaluated just on the basis of technical factors: social factors are of equal importance.

But good technology is important in the long run in improving social conditions. Therefore, if social factors permit, methods which give higher yields of better quality charcoal at lower cost should be used.

Technical equipment appropriate to industrial charcoal production on a small scale has come into use only since 1965. Since then, these new developments have followed an ever-increasing upward trend, not previously known in the charcoal industry.

It may come as a surprise to the outsider that the investment costs are frequently a minor concern of the entrepreneur because the adaptability of modern equipment allows the investor to reduce his risks considerably and to add new capital equipment whenever his funding is sufficient and the market demands it.

7.3 Conclusions

Charcoal is a growth fuel belonging to the class of energy sources that is growing rapidly; namely, the supplementary fuels for household and industry. It was once mankind's first and only fuel for the extraction of iron and other metals from their ores and paved the way to industrial development all over the world.

Charcoal and its derivatives sparked the start and rise of the chemical industry a hundred years ago. Thereafter, the use of charcoal declined to a low level in industrialized countries, where more concentrated forms of energy were available. But it never ceased to be important as an industrial and household fuel in developing countries.

Charcoal and its by-products will increase their contribution because of the synergistic effect of improved charcoal technology and advanced silvicultural practices.

It must be admitted that the capital outlays for improved charcoal equipment, large or small, appear at first sight to be a barrier to their use in developing countries. However, a closer analysis and comparison shows that in terms of capital cost per ton of charcoal produced, over the life of the plant concerned, the differences are not as marked as is frequently thought.

Where the capital outlay for the investment is high, it is usually compensated by considerable savings in labour wages, lower costs for the utilities, and higher profitability through upgrading the quality of the product.

Much of the abundant resources of forestal waste and agricultural residues is at present untapped, burnt away or wasted owing to the lack of technical expertise and economic encouragement by governments. Because these energy resources are renewable and their carbonization can help to optimize the overall utilization of forests and farm crops, the new charcoal technology has a special significance for all developing countries which hold these large reserves and need additional funds to practice economically.

What should governments do?

- Either within the Planning Department or in the Forestry Department, a section should be set up concerned with the promotion of a well-organized industrial charcoal production.

- Guidelines or legislative measures should be issued to define and clarify the place of the charcoal industry within the existing energy concept.

- Since modern charcoal technology draws on biomass resources provided by forests, industry and agriculture, administrative roadblocks between competing agencies must be removed to give all concerned a clear view of the national concept.

- Eventually, the government holds the key to the success or fiasco of resource preservation programmes. Therefore, the administration must clearly state priorities for the use of national reserves from which the fuel for households and industry will be made in the future.

Further recommendations are:

- A pilot demonstration project should be implemented to investigate the validity of planned ventures and to study operating characteristics.

- Training programmes should be conducted.

- Data should be collected on resources, markets and industrial applications for products of the charcoal industry.

-   Information should be disseminated to interested groups, universities, and private investors.

-   A study should be made of the possibilities of local equipment suppliers and their potential to provide the necessary equipment from indigenous maintenance shops.

Industrial charcoal-making and silviculture go well together and can be instrumental in preventing resource depletion. But much work remains to be done. The first priority is to spread information on the capability of modern charcoal-making to utilize waste materials.

References

(1) F. Fluegge, Chemische Technologie des Holzes, (56, 57), Munich FRG, 1954

Chapter 8
CHARCOAL BRIQUETTES AND ACTIVATED CHARCOAL MANUFACTURING

There is always a need to make charcoal more convenient in use, and special industrial applications require charcoal in agglomerated shapes. For this purpose, several techniques are available to the charcoal producer: extrusion, pelletizing and briquetting.

The latter is by far the most common method. Briquetting units can be attached to the carbonization plant using excess heating gas for the briquetting dryer. They can be designed economically, starting from a capacity of a few hundred tons per year. The largest known briquetting operations turn out between 70,000 and 80,000 tons annually, employing automatic equipment for pressing, drying and bagging.

The term "briquettes" is normally applied to the material obtained by thoroughly mixing powdered charcoal with a binder. This is a thick liquid consisting of starch, molasses, tar, etc.

Further additives are inorganic fillers which delay the heat release of the burning briquette, and ignition enhancers (sodium nitrate) to make them "easy-lighting".

In any event, charcoal briquettes cannot be lit by simply putting a match to them. The usual kindling materials are paper or the special charcoal lighters (solid and liquid) sold in stores.

8.1 The Briquetting Process (1)

The equipment described here is suitable for a medium-sized plant with a briquetting capacity of 1,000 to 5,000 tons/y.

Preparation: The raw material is received in a hopper and an accurate feeder accumulates the charcoal and feeds it to the pulverizer. The total plant output is centred in the feeder and pulverizer.

Crushing: Usually a hammer mill is installed to size the material to the desired screen analysis. The screen analysis will depend very much on the type of briquette to be produced.

Mixing: A mixer which provides the extra retention time to guarantee complete blending with the binder, filler, additives, etc. is necessary. Thorough mixing can also reduce the quantity of binder considerably.

Forming: The most commonly used presses are the roller types that may be adjusted to differnt pressures and speeds. Photo 14 shows a roller press discharging pillow briquettes and Photo 15 a close-up of the briquettes.

Selection of binder: The most common binder is corn starch. As well as all types of starch, sugar cane molasses can also be applied. Tests will show which type of binder is best suited and will also determine the composition. Much attention should be paid to this question, because the cost of the binder will contribute significantly to the total production cost of the briquettes. The briquettes should be subjected to a burning test. Odour and visible smoke caused by some binders are undesirable. Also the briquettes must be resistant to any fermentation agression, at least for 18 months (2).

Selection of filler: This additive consists of an incombustible mineral. In most cases, limestone is used, but other materials such as ground oyster shells are also frequently found in briquettes. The addition of a filler

PHOTO 14   A charcoal briquetting press. Two
              rows of moulds can be seen and the
              wire belt.
              (Photo W. Emrich)

PHOTO 15   Pillow-shaped charcoal briquettes.
                (Photo W. Emrich)

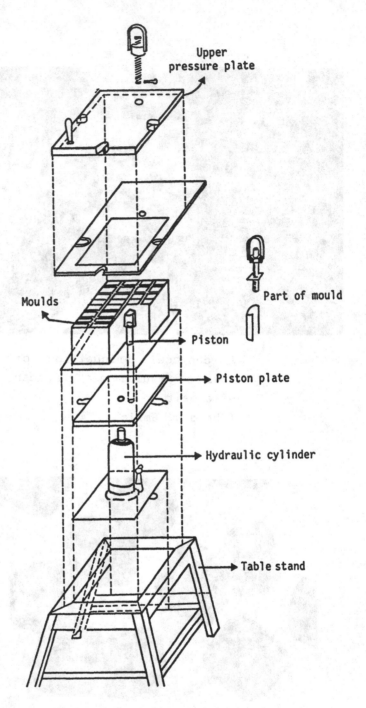

Upper pressure plate

Part of mould

Moulds

Piston

Piston plate

Hydraulic cylinder

Table stand

Figure 50. Simple Charcoal Briquetting Press
(Courtesy of Carbon International, Ltd.,
Neu-Isenburg, F.R.G.)

serves two purposes: prolongation of the cooking time by slow heat release (bakeoven effect); and reduction of the calorific value of the briquette (fuel-saving effect).

A well balanced briquette composition will display superior burning characteristics, particularly cooking time, when compared with lump charcoal.

Sand and high silica-containing fillers are not suitable as energy extenders because of their abrasive properties; they will wear out the moulds of the briquette press in a short period of time.

Additives: The briquetting industry has found numerous additives during the last 10 years which will enhance the ignition characteristics, prevent fermentation, or accelerate chemical processes when charcoal is involved. Other additives serve decorative purposes only, i.e. to give a certain flame colour.

8.1.1 Simple Briquetting Equipment

Frequently the need for low cost equipment arises especially in small charcoal operations and where the produce is to be dumped into markets not demanding high grade charcoal briquettes. In all these situations the charcoal-maker likes to turn to equipment which can be assembled easily by local workshops.

If the capital outlays for labour do not bear great weight on the economics of the venture one may prefer more manually operated apparatus and paraphernalias also.

In general the assemblage of the machinery should leave to the charcoal-maker the possibility to add more when his production needs expansion and his budget can afford it. This, of course, implies the idea to employ smaller machines and an outfit for which the constituent elements are at hand when they are wanted.

Briquetting press

Usable are all kinds of presses which will permit satisfactory solidifying of the mixture, which normally can be achieved by a raised pressure. Anyway, tests will give the right answer.

There are numerous types of presses of which charcoal-makers operate modifications: piston presses, table presses, tablet presses, soap and stamp presses.

Figure 50 displays a design which is commonly in use and can be fit together by any workshops equipped with basic tools. The illustration is self-explanatory. Up to a limited reach the shape of the charcoal agglomerates may be altered also.

Drying equipment

For the drying of the "green" briquettes two methods have been commonly adopted: stationary drying chambers or cupboards and mobile drying trays or racks with wire cloth.

The latter way of desiccating is more popular in medium sized charcoal operations. The trays with the clustered charcoal are placed on stands which can be wheeled into the drying kiln, which usually is a masonry structure.

It is normal practice to heat the kiln with hot gases obtained from the burning of charcoal gas in an adjacent fire chamber.

8.2 The Activated Charcoal Process

Although unactivated charcoal already has some "active properties", it would not meet present-day requirements. To improve and enlarge the active surface of charcoal, two processes are mainly used: gas (steam) activation, and chemical activation (3).

Gas (Steam) activation

Here one starts from pre-carbonized material which has not usually been manufactured or mined for the purpose of activation, e.g., wood charcoal, coconut charcoal, peat charcoal, brown-coal coke, or even coal or the like.

The raw material, in the form of lumps or finely ground, is subjected to the action of gases such as water vapour, carbon dioxide, air, or mixtures of them, at 700 - 1000° C. Oxygen present in the free state, or combined in the gases, burns up carbon and produces the desired pores. Also the powdered carbon is often extruded in presses before activation, with the aid of coking binders. For example, in the production of pellets, powdered wood charcoal is intimately mixed with hot coal-tar pitch and with activation accelerators and the mixture is extruded through dies. The axtruded product is carbonized at only moderate temperatures. The baked grains, reduced in size by handling operations, are however still inactive. They are finally activated by water vapour in a rotary furnace or in a fluidized bed at 700 - 900° C. The water vapour removes carbon from the interior of the grains in accordance with the equation:

$$C + H_2O = H_2 + CO$$

and thus produces the desired porous structure.

Kilns, rotary furnaces, Herreshoff furnaces, and fluidized bed furnaces are all used. The possible variations in procedure, taken in conjunction with the choice of raw material, provide an almost unlimited number of types of activated carbon having different absorption properties. The art of the processor lies in conducting the activation process in such a way that combustion of the carbon does not take place from the outside of the grains. Products with consistent properties can be produced only when the production conditions are accurately known and strictly adhered to. This applies to all activation processes.

Figure 51 shows an activation plant for the production of activated carbon pellets (see also Photo 12).

If the aim is to produce a granulated or powdered product, the machinery or equipment required will be simpler.

Chemical activation

Chemical activation generally starts with uncarbonized vegetable products, such as sawdust, peat, etc. These are impregnated with zinc chloride solution or phosphoric acid, to name the most commonly used materials. If the carbon product is to be in the form of a powder, it is calcined directly, at 400° to 700° C, after previous drying. For granular carbons, one can start from granular raw materials. However, better activated products with harder grains result if the "green" mix is made from finely divided material and the mixture is extruded to give moulded forms. These moulded forms must be thoroughly dried before calcination. On heating (calcination), the added chemicals draw water out of the raw materials and result in a porous carbon structure. The carbon is dried after washing out the activating chemicals. Thus, the chemical additives are recovered and returned to the cycle.

As in gas activation, a large number of different types of activated carbon can be produced by varying the characteristics of the green mixture, by special additives, and by the way in which calcination is carried out.

There are many patented processes for the production of activated carbon, but none are followed rigidly by current carbon producers. In fact, most processes are unpatented internally developed techniques. The competitive nature of the activated carbon industry has required carbon manufacturers to keep details of their processing techniques quite secret.

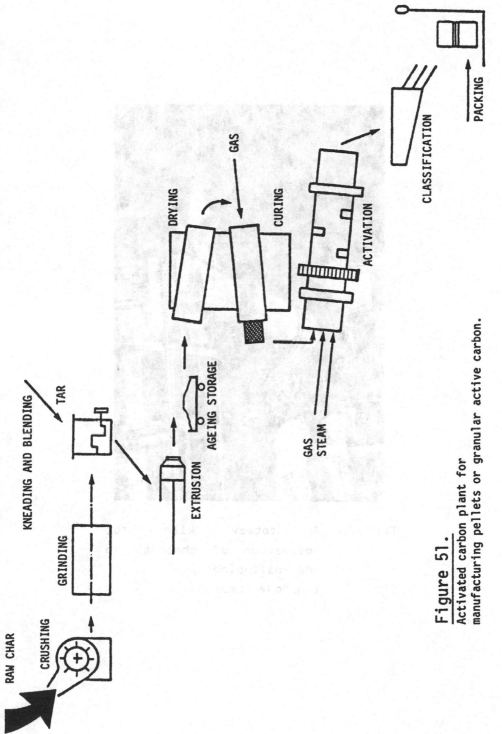

RAW CHAR

CRUSHING

GRINDING

KNEADING AND BLENDING

TAR

EXTRUSION

AGEING STORAGE

DRYING

GAS

CURING

GAS

STEAM

ACTIVATION

CLASSIFICATION

PACKING

Figure 51.
Activated carbon plant for manufacturing pellets or granular active carbon.

Photo 16 A   rotary   kiln   for
activation of charcoal in
the Philippines
(Photo W. Emrich)

Although details are not available, it is certain that most producers follow rather standard steam activation methods with very close control. Major processing variations arise from adapting to different source materials and from pre-activation and post-activation carbon treatment designed to yield particular properties for carbon suitable for specific applications (4).

To service the activated carbon market adequately, specific requirements must be met. To enable the charcoal producer to meet these requirements, each production process must be investigated to determine the best method of making the desired products.

REFERENCES

(1) Y. Yoshida, Status of Hot Briquetting and Form Coke Technology in Japan, 1971.

(2) IBA Proceedings, The Institute for Briquetting and Agglomeration, 14th Biennial Conf., Reno, Nevada, 1981.

(3) M. Smisek, Active Carbon manufacture, Properties and Applications, Elsevier, Amsterdam, Netherlands 1970.

(4) H. V. Kienle, Aktivkohle und ihre industrielle Anwendung, Stuttgart, FRG 1980.

Chapter 9
SAFETY PRECAUTIONS AND ENVIRONMENTAL CONSIDERATIONS

9.1 Safety in Charcoal Operations

Whether charcoal is made in the traditional way or by industrial methods, two hazards are always present: explosions of gases and dust, and fires in the stored charcoal.

Accidents can be greatly reduced by making use of safety features and adopting safe work habits. In all plants where high-temperature operating conditions are commonplace and extensive, carelessness or slovenliness can be ruinous. Production hazards increase, and undesirable or dangerous burning conditions can arise if the operator neglects to pay close attention to such vital operational factors as the course of the converter temperature, pressure indicators, structural conditions of the production equipment and storage bins. There are ample records of plant damage and destruction caused by such neglect and oversight.

Explosions

The causes of such accidents are not clearly understood. They are thought to be caused most often by a mixture of retort or converter gases with air.

In batchwise operated retort operations, the greatest quantity of gas is present in the retort at the end of the coaling time, hence the admittance of more air than is needed to run the process can form highly explosive gas mixtures. In a continuously run converter, the admittance of more air than is needed may cause over-temperature in the first place and form explosive mixtures in the off-gas system. However, according to Swedish sources, frequent "puffings" have been observed during the early stage of coaling, when comparatively large volumes of water and other vapours are being condensed on relatively cool raw material. The non-condensable gases, including of those capable of explosion, are thus free to form critical mixtures with air.

Fires

Explosions are a major cause of fires. Whilst the explosion itself may cause only minor damage to the system, fire can result from admission of large quantities of air to the retort, converter or off-gas system through cracks.

Other causes of fire are the operator's unfamiliarity with proper operating procedures and outright carelessness. In high-temperature operations, there is always the danger of wall separation. Should accidental openings occur and remain unnoticed, the seepage of excessive amounts of air through them could easily change the temperature pattern. This may result gradually in very high temperature outbreaks, or they might take place very rapidly, creating a serious fire condition. The operator's familiarity with his equipment and the steps necessary for counter-action are the best insurance for safe practice and satisfactory production. Well established, periodic inspection of the industrial charcoal-making plant will often indicate the corrective measures necessary for proper control and reduce the possibility of damaging fires.

Also of major importance are the reduced yields and loss of operating time caused by improper sealing or structural leakages of air in the charcoal cooling bins during the cooling period. Such conditions may occur even when a well-standardized pattern of operation has been established. The importance of inspecting and maintaining cooling bins during the cooling cycles, controlling operational conditions, and using safe practices cannot be over-emphasized.

Skin irritations

The tars and smoke produced by carbonization, although not directly poisonous, may have long-term damaging effects on the respiratory system.

Wood tars and pyrolytic acid can irritate the skin and care should be taken to avoid prolonged skin contact by providing protective clothing and adopting working procedures which minimize exposure.

Hazards to the public

Fire, whether controlled inside the retorts or converter or uncontrolled, constitutes a potential hazard for the public. Unauthorized persons, including the public, should not be admitted to the plant unless guided. Safety helmets are a necessity for the workmen and the visitor. The transport of wood or any raw material, charcoal handling and other essential work involves hazards. Therefore, safety measures and safe work habits are of prime importance.

9.2 Safety Devices and Equipment

Pressure-relief doors

Explosions are always a potential danger when handling a dusty material, or one which contains gaseous vapours. Feed and storage bins are therefore designed with explosion relief tops which lift to vent the gases when the internal pressure rises abnormally, e.g., above 350 - 400 mm water. In addition, dead-weight relief doors are usually incorporated in the tops of the bins. These doors should lift at a lower internal pressure.

Automatic temperature shutdown

In the event that temperatures inside the retorts, converter, or the off-gas system exceed predetermined limits, the air supply, the heating fans of the retorts or the fan of the converter gas system are shut down. At the same time, the air supply of continuously operated converters is cut off and the converter gated off. Residual gases are vented through the emergency flare.

Electric power failure devices

If a total power failure occurs, the air supply and the draught fan stop and all shut-off gates remain in prefailure condition. When the process becomes dormant, the system is gated off.

Temperature indication and control

The equipment that indicates the operation temperatures and controls the safety devices must be selected very carefully. Regular inspection and proper maintenance are a necessity.

9.3 General Safeguarding of Charcoal Plants

Water supply

A water supply is highly important to any charcoal plant. A hose with a nozzle should be kept ready for immediate use at assigned points in the plant. Back-pack water pumps or large-capacity fire extinguishers provide some measure of fire protection.

Detection of poisonous gases

Poisonous gases (carbon monoxide) are present in retorts that have just been cooled down, or in the off-gas system just after shut-off. There are portable detectors on the market which indicate the concentration of the gas confined in the production apparatus.

However, the best prevention is to ensure thorough ventilation before workmen or maintenance staff enter the bins or retorts and also during the entire time they are occupied.

Safety manual

All safety instructions and any changes in them must be made known to the workmen without delay. The time-proven practice in all well-organized plants is that every workman has to attend an additional instruction lesson each quarter, given by the plant engineer. But common sense remains the most important factor in preventing and if necessary coping with hazardous situations in the plant.

First-aid accessories

Adequate first-aid supplies, including dust and gas (carbon monoxide) masks, should be kept available at the central point.

9.4 Precautions for Charcoal Storage

Customarily, the charcoal demand is seasonal which makes it necessary to stockpile a considerable inventory.

A great deal of care must be taken in storing freshly produced charcoal. It has a tendency to absorb oxygen from the ambient air. Rapid absorption, however, generates considerable heat which builds up to a point where the stockpiled char starts to burn.

Tightly packed masses of charcoal fines and charcoal with a high content of volatiles are more subject to spontaneous combustion than the larger lump charcoal. Self-ignition may occur if charcoal has been water-sprayed for better cooling.

It is, therefore, advisable to place freshly discharged charcoal in the open, separated from previously cooled and conditioned charcoal, for at least 24 hours. During this time, the char should be exposed to air circulation and protected from rain and wind, preferably in an open shed rather than under a tarpulin. If there is no evidence of heat or active fire after the 24-hour period, the charcoal may be considered safe for warehousing.

9.5 Environmental Considerations for the Charcoal-Maker

During the last two decades, environmental control and the related legislation have become important concerns of charcoal-makers. Numerous cases are known where plants under operation for more than half a century have had to be shut down or radically changed as a result of legislative pressure. The potential charcoal-maker should take these aspects into considerations before spending money and setting up a plant.

Fortunately, the feed input of an industrial charcoal plant is very largely removed in a solid stream as charcoal, and the ejected off-gases are significantly reduced by the removal of the condensate, which results also in a much cleaner off-gas stream as compared with traditional charcoal-making.

The combustible gas generated by the system will burn cleanly in a stack if not otherwise utilized.

Another favourable feature of industrial charcoal-making is the fact that these systems discharge minimal liquid effluents. Wastewater does not usually occur except in plants where the by-products are fractionated from the original pyrolytic oil by distillation.

The environmental aspects of industrial charcoal-making will depend very much on the type of raw material and also the throughput capacity. Therefore, component effluents can be discussed only in a general way.

Raw material preparation

Hogging machinery usually causes noise levels which may exceed local decibel allowances. Normally, the noise is confined by enclosing the machines, and in residential areas the machinery is not operated during regular night shift.

Dryer exhausts are dust-loaded and the degree is related to the size of the feed (fines). Cyclonic equipment is installed to depress the grade of dust exhaust and serves at the same time to prevent undesirable losses of raw material by retrieving it.

Retorts and converter

Any fugitive vapours or dust from these units occur only when they are opened. In continuous operations, a retort or converter will only be opened during the shut-down period and therefore, no preventative measures are needed.

Char handling

Conveyer belts are normally enclosed. The char bin's outbreathing is passed through a bag collector to retain the char dust. Captured dust will be periodically vibrated from bags and allowed to settle inside the bins.

Scrubbing system for pyrolytic oil

Any fugitive vapours can be collected in draughted vent hoods. They are burned in an existing combustion device (flare stack).

Emergency venting

The systems are designed so that in the event of an unusual pressure rise within the system or a fire, the vapours are vented to the atmosphere. This is done for personnel and equipment safety. On the basis of long experience, it can be stated that such occurences are uncommon, and when they do occur they last only a few seconds. Since the systems are usually outside installations, the vapours disperse quickly.

Wastewater

Industrial charcoal-making plants discharge very little water to sewage. Water is used primarily for cooling and is recirculated in closed-loop systems. The primary water effluent is from the distillation systems for the fractionation of pyrolytic oil. In this case, the water will contain some organic liquid and it should be sent to a chemical after-treatment stage.

The tars and pyroligneous liquors can seriously contaminate streams and affect drinking water supplies for humane and animals. Fish may also be adversely affected. Liquid effluents and wastewater from medium and large-scale charcoal operations should be trapped in large settling ponds and allowed to evaporate so that this water does not pass into the local drainage system and contaminate streams.

Kilns and pits, as distinct from retorts and other systems, do not normally produce liquid effluent because the by-products are mostly dispersed into the air as vapours.

Precautions against air-borne contamination of the environment are of greater importance in this case. Therefore, kiln batteries may not be set up in the neighbourhood of residential areas since their smoke emission would cause a nuisance. Smoke emission will also be the limiting factor for the number of kilns assembled in a battery.

Chapter 10

CHARCOAL LABORATORY WORK

Whereas the traditional charcoal-maker rarely engages in analytical work, the industrial charcoal producer cannot do without it. Raw material composition, quality control, investigation of customer complaints, environmental and safety regulation, etc. are the problems to be dealt with. Fortunately, the analytical procedures to be performed do not require extremely sophisticated and expensive equipment. The procedures for analysis are the same as those employed for bituminous coal and these are set out in national standards (e.g., for Germany the relevant standards are contained in DIN 51749). The cost of equipping a charcoal laboratory with a basic inventory is around U.S. $ 8,000. This figure does not include the provision of space and furniture.

The laboratory staff must have a certain degree of skill and training. In most plants, one or two technicians and two assistants are on duty during normal operational times, which means also during night shifts. The following is a compilation of the most common test and analytical procedures. Although most of them were developed many years ago, and have been commonly adopted by the charcoal industry, some suppliers still use their own methods. Therefore, in comparing analytical results, one has to know the procedure by which they were obtained.

Whether raw material or carbonization products have to be analysed or tested, the sampling is important and must be done carefully. In the case of charcoal, the collected samples of a batch or of a truckload ready for shipment are best mixed in a drum which is filled to only half its capacity. After rotating the drum for several minutes, a sample is taken off and ground to a mesh size of under 1 mm.

The laboratory takes 200 g of the sample. Half of it is set aside for storage in a closed and sealed tin can. This sample serves as proof for further investigations and discussions, if they become necessary (1).

10.1 Analysis

- in raw materials:

Accurately weigh 10 g of the crushed or ground sample and dry it in an electric drying chamber at a constant temperature of 105° C. Weigh it after 3 hours and continue the weighing and drying at 24-hour intervals until the loss is not more than 0.25 % in day's drying time. The loss of weight is calculated as a percentage of the initial wet weight.

In the case of wood logs, blocks must be cut with an approximate size of 5 x 5 x 6 cm. The procedure for drying and determination of moisture percentage are the same as for crushed and ground material.

- in charcoal and charcoal briquettes:

Crush the sample in a porcelain mortar with a pestle and weigh accurately 3 g. Drying and weighing procedures are the same as above.

Ashes

- in raw material:

Crush or grind the sample and weigh accurately 3 g in a platinum or porcelain crucible with a lid. Heat up to 700 - 800° C in an electric muffle oven. Check by weighing and heating at intervals until the weight loss is under 0.25 % of the initial weight.

The difference between dry initial weight and weight of the sample after combustion of all organic matter is expressed as a percentage of initial weight, and is ash content.

- in charcoal and charcoal briquettes:

Prepare the sample as above. According to estimated ash content, accurately weigh 3 to 5 g in a platinum or porcelain crucible with a lid. Then follow the procedure described above.

If the charcoal does not burn completely, apply several drops of hydrogen peroxide (3 % solution) and heat the crucible, supported by a wire triangle on a tripod, with the flame of a bunsen burner.

Volatiles and Fixed Carbon

This method is applied to all charcoal products: lump charcoal, charcoal fines, granules, pellets and charcoal briquettes. By heating the charcoal under the exclusion of air (oxygen), the confined gases are expelled. These gases are commonly termed "volatiles".

- preparation of charcoal sample:

The necessary crushing should be done manually in a porcelain mortar with a pestle. Forced grinding is not recommended because of the heat which could be generated and which would drive off part of the volatiles. Dry the sample at 105° C (not higher!)

- determination of volatiles and fixed carbon:

Accurately weigh 1 g of the dry powdered sample in a platinum (preferably) or porcelain crucible with a lid.

Dimensions of the crucible are: lower diameter 22 mm, upper diameter 35 mm, height without lid 40 mm. The lid has a pinhole in the centre with a diameter of 1.5 mm (not wider!). The rim of the lid must comfortably overlap the brim of the crucible to prevent the intake of air during heating.

Stand the crucible, with a wire triangle, firmly on a tripod and heat the bottom gently with a shining flame. The distance between bunsen burner and crucible should not be less than 6 cm. After 2 - 3 minutes, open the gas adjusting screw and the air control of the bunsen burner to full capacity and continue until the small flame above the pinhole in the lid goes out. This indicates that all volatile matter has been driven off (1).

Put the hot crucible into a desiccator with calcium cloride in the bottom as a desiccant and leave until the sample has cooled off. Weigh it as usual.

The difference between the initial weight and the final weight is the content of volatiles.

The value for fixed carbon is calculated by the following formula:

$$C_{fix} = 100\ \% - (\text{volatiles} + \text{ashes})$$

Sulphur

This is usually evaluated in all charcoal products. The most common method is calorimetric combustion of the charcoal with the addition of dilute sodium hydroxide. The formation of sulphates permits precipitation with barium chloride as barium sulphate.

Accurately weigh 1 g of the dry powdered sample and insert it into the calorimeter according to the instructions given by the supplier of the equipment. Combustion takes place under an excess of oxygen, and the pressure is kept at 20 atm. After ignition, the calorimeter must be shaken for about half an hour (1).

Then release the pressure and discharge the combustion residures into a beaker by several rinsings with water (the contents of the calorimeter must be completely transferred).

Heat the beaker to boiling point and add 10 cm^3 of barium chloride solution (approx. 250 g $BaCl_2$/1000 cm^3 distilled water). Immediately a white precipitation of barium sulphate will appear.

After cooling off, the contents of the beaker are poured through a glass filter which collects the barium sulphate precipitate. After several rinsings with distilled water, the filter is dried and weighed to determine the content of barium sulphates. This is the difference between the weight of the empty filter and the weight of the carefully dried filter with the precipitate in a drying oven (105° C).

Use the following formula to calculate the sulphur in the sample (accurate 1.0 g):

$$\% \text{ sulphur} = \frac{\text{weight of dry barium sulphate} \times 32 \times 100}{233.5}$$

Screening analysis

Charcoal fines and charcoal powder are classified by the percentage distribution of grain sizes. For the analysis, a set of standardized screens are used consisting of several screens with different mesh sizes. Most common is the following set: (2)

No.	Mesh size	Wire thickness
4	3.3 mm	1.0 mm
3	2.5	1.0
2	1.5	1.0
1	1.0	0.65

Assemble the screening set in order, with screen no. 4 on top and no. 1 in the lowest position. Accurately weigh 100 g of dried sample and put it in screen no. 1.

Shake the whole csreening set for 2 minutes, making two to-and-fro movements per second. Then weigh the residues of the fines remaining on each screen. Note down in your report the different weights as percentages of total initial weight in the following order:

```
grain size   IV  =  3.3 mm       (screen no. 4)
   "    "    III  =  2.5 - 3.3 mm (screen no. 3)
   "    "     II  =  1.5 - 2.5 mm (screen no. 2)
   "    "      I  =  1.0 - 1.5 mm (screen no. 1)
   "    "      0  =  screen transit
```

For the classification of lump charcoal, larger mesh sizes are necessary to evaluate the grain size distribution which normally ranges between 10 and 120 mm.

The friability test

Friability tests are a means of measuring the tendency of the charcoal to break into smaller pieces when subjected to repeated handling, and so indicate the relative extent to which size coals will decrease in size during transport, or descent inside a blast furnace.

The figures in % indicate the reduction in size which the charcoal has suffered during the test. Therefore, the lower the % figure, the stronger is the charcoal.

- evaluation of friability by the tumbler test

This is considered the most important of the friability tests. It is derived from Recommendation R-556 on the MICUM indices of coke of the International Organization for Standardization (ISO).

10 kg of charcoal are placed in a steel test drum 1,000 mm long, 1,000 mm diameter, fitted with four steel angles fixed lengthwise inside the drum. The drum is rotated at 24 rpm for one hour (total of 1,440 revolutions).

Charcoal is first tested at the works reception. The coal is hand-sieved and only material larger than 31.75 mm is tumbler tested to avoid the possibility of fine sizes protecting the coarser pieces by their cushion effect against shocks and abrasion.

Before the tumbler test, the new average size of the charcoal must be calculated on the results of the screening analysis.

Bulk density of charcoal fines

This indicates the weight of the charcoal fines per unit volume and is an important datum for shipments.

Pour the charcoal sample, as it is received from the plant or storage in three discrete portions into a calibrated cylinder, one at a time. After filling in each portion of the sample, stamp the cylinder vigorously on a wooden board to the point where further stamping does not reduce the volume of the cylinder content. When the calibration mark of 100 cm^3 is reached, stop and weigh the charcoal fines.

The obtained weight multiplied by 10 gives the bulk density per litre.

The procedure may be facilitated by employing a shaking machine.

Viscosity of pyrolysis oil

For the measurement of viscosity, several standard apparatuses are on the market. They all operate according to the general principle of comparison of the sample with a liquid of known viscosity.

Most common in the charcoal industry is the determination according to the Engler scale. The instruction comes with the apparatus.

Flash point of pyrolysis oil

This is the lowest temperature at wich a fuel in an open vessel gives off enough combustible vapour to produce a momentary flash or fire when a small flame is passed near its surface. Special apparatus can be purchased.

Calorific value

This is the number of heat units obtained by the complete combustion of charcoal, charcoal fines, charcoal briquettes, pyrolysis oil or off-gas. For the determination, an oxygen calorimetric bomb is necessary, which can be purchased. Follow the instruction manual.

Sampling of gas

For gas sampling, some general rules must be applied to avoid inaccuracies and mistakes. The off-gas sample should be tested immediately, otherwise alterations of its composition may take place. If testing is not carried out immediately, the gas sample may be stored only in a glass flask or rubber container. All containers must be sealed tightly after sampling.

Besides its calorific value, the composition of the gas is also important for the plant operation. The composition can be determined by suction of the gas through specific absorption agents which retain one gas component but let others pass. These apparatuses can also be purchased.

General remarks

The above enumeration of analytical procedures is far from complete. There are many other operations which will be required only once a year or even less frequently.

Some items of the laboratory work described in this handbook may also be rarely required in practice, particularly in a small plant which produces two commercial products, as happens frequently.

A limiting factor may be that a skilled technician is not available, or proper laboratory equipment is not at hand. However, it has always been the practice of small-scale charcoal producers to engage the help of an existing laboratory which could be either a facility of the university or private. Normally, the charcoal beginner quickly learns what he must do himself and which part of the laboratory work should be contracted out.

In all cases known to the author, the personnel question has been resolved by on-the-spot training of an adaptable person.

A charcoal-maker engaging in an export business usually has to rely on his partners abroad, who should be well equipped and have capable staff to deal with the problems that arise and to give useful advice to their suppliers.

10.2 Bench-Scale Carbonization Tests

For the charcoal industry developer as well as for the plant manager, it is indispensable to be familiar with the ways to conduct carbonization tests. In practice, there are many reasons for: investigating new raw material, checking the yield efficiency of an existing charcoal plant, or just for demonstration purposes.

The essential parts of the apparatus are shown in Figure 52 which is self-explanatory.

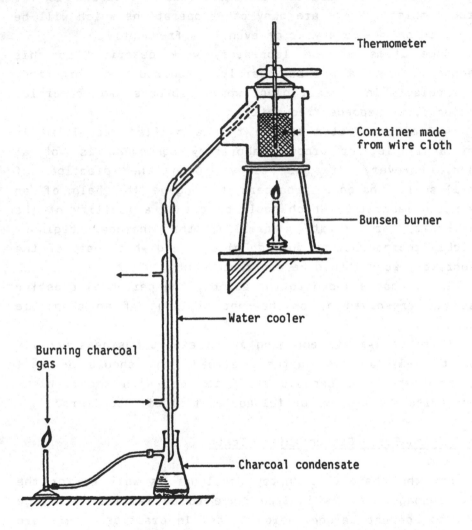

Figure 52. Apparatus for bench-scale dry distillation.

Thermometer

Container made
from wire cloth

Bunsen burner

Water cooler

Burning charcoal
gas

Charcoal condensate

The small retort has a capacity of 1,000 cm^3 and can be made from copper, aluminum or stainless steel. The thermometer must be calibrated above 550° C.

The retort should be filled to about three-quarters full with the prepared raw material sample and the lid closed airtigth. Heating with the open flame of the bunsen burner should start slowly and proceed according to the description in Section 10.1.

After the terminal temperature has been reached, the obtained products - charcoal and pyrolysis oil - can be weighed to determine the yields.

It is also possible to obtain data for off-gas with the apparatus.

It should be borne in mind that this laboratory retort indicates data for intermittent carbonization processes only, and they cannot be applied to continuous projects in a straight line. But the results will give valuable information and insight for planning and decision-making.

References

(1) G. Bugge, Neue Untersuchungsmethoden fuer die Produkte der HIAG-Werke, 1947, (private paper).

(2) SABS 1399 - 1983, SA Bureau of Standards, Pretoria, S.A.

A P P E N D I C E S

Appendix 1 C A S E   S T U D I E S

Charcoal plant with continuous C.I.S.R. Lambiotte Retort

Annual production of lump charcoal   2,200 t
Usable calorific excess energy 36×10^9 KJ

Raw material: 10,000 t of dry wood (30 % moisture)

Charcoal plant with continuous Vertical Flow Converter (Tech-Air)

Annual production of charcoal briquettes   4,500 t

Pyrolysis oil   2,800 t

Calorific excess energy used for
drying of feedstock

Raw material: 12,000 t of sawmill and
agricultural wastes (air
dry)

Note: The capacities of both plants are adaptable over a
wide range and can be modified.

CISR LAMBIOTTE RETORT[x]

I. PLANT CAPACITY

The suggested plant comprises one or more continuous C.I.S.R. LAMBIOTTE RETORTS, 2200 t charcoal per year per unit.

This medium capacity has been chosen to avoid a high transport cost for the wood but at the same time, to benefit from a high technological level, i.e., high output, productivity and quality.

Annual wood consumption per unit: 6600 t of dry matter -- 10,000 t of 34 % moisture wood.

Annual charcoal production: 2200 t (350 working days).

Energy balance of the process:

a) Input 2.77×10^{10} kcal = 100 %
    (4,200 kcal)kg)

b) Output
    2,200 t lump char- 1.49×10^{10} kcal = 53,8 %
    coal (6,800 kcal/kg)

c) Process losses
    3,400 t of moisture
    to be evaporated 0.34×10^{10} kcal = 12.3 %
    (1,000 kcal/kg)

    3 % heat loss
    through insulation 0.08×10^{10} kcal = 3.0 %

    Remaining energy 0.86×10^{10} kcal = 30.9 %.

x) Information provided by the manufacturer

The remaining process energy (excess energy) is equivalent to the calorific value of 800 to 900 t of fuel oil per year.

II. COST CALCULATIONS

A. BASIS
- Free wastes of a sawmill (slabs) 10,000 t per year at 30 % moisture.

- Manpowever: 7 people. Average cost $ 7,000/yr per man.

- Charcoal selling price: $ 0,25/kg.[x)] 2200 t per year production.

- Capital: a) entirely loaned (10 % year)
   b) entirely subscribed

- Total investment: $ 732,000.

- Amortization: 15 years, $ 48,800/year.

- Delivery and erection delay: 6 months.

B. Calculations

Annual cost:	Manpower	$ 49,000
	Others (water, electri- city, bags, maintenance, etc.)	$ 47,000
Loan interest:		$ 73,200
	subtotal	$ 169,200
Amortization		$ 48,800
	Total	$ 218,000

x) Average market price for lump charcoal in European markets, 1981.

Details of the $ 47,000 "others" cost:

Gazoil, grease	$	5,000
Electricity		12,000
Bags (25 kg)		18,000
Spare parts		6,000
Maintenance		6,000

If the capital is not loaned but is entirely subscribed, the situation becomes:

1st year: sales $ 275,000
               charges   144,800
               (manpower, others,
               amortization)   _____

               Profit $ 130,200
               (taxes excluded)

2nd year: sales $ 550,000
               charges   144,800

               Profit $ 405,200

3rd year: sales $ 550,000
               charges   144,800

               Profit $ 405,200

At the end of the second year, the return on the capital is 50 % per year before taxation.

VERTICAL FLOW CONVERTER (TECH-AIR)[x]

I. PLANT CAPACITY

The proposed plant is composed of one vertical flow converter (medium size) with a complete pyrolysis oil recovery unit. A briquetting machine, briquetting dryer and bagging equipment are attached to the charcoal production.

The capacity of the plant can be adapted to small operations utilizing the wastes of sawmills, nut processors, copra dryers, sugar factories, and plantations, etc.

Raw material consumption

Types of raw materials: Sawmill waste, residues from forests, brushes, leaves, nutshells, cotton sticks, sugarcane bagasse, bark, coffee husk (pulpa), municipal waste, etc.

Max. length of particles 50 mm, diameter 7 mm.

Annual consumption 10,000 t (dry material) = 11,000 t (airdry material).

Annual production

The plant can be run with or without pyrolysis oil recovery. If no pyrolysis oil is produced, the charcoal yield will be significantly higher:

charcoal briquettes	4,500 t/y	or	6,000 t/y
pyrolysis oil	2,800 t/y	or	none
heating gas	none		see energy balance

---

x) Information provided by the manufacturer

Energy balance of the process

Input

10,000 t dry biomass (4,200 kcal/kg)	$4,20 \times 10^{10}$ kcal =	100.0 %

Output

4,500 t charcoal briquettes (6,000 kcal/kg)	2.70×10^{10} kcal =	64.3 %
2,800 t pyrolysis oil (4,000 kcal/kg)	1.12×10^{10} kcal =	26.7 %

Process losses

1,000 t of moisture to be evaporated (1,000 kcal/kg)	0.10×10^{10} kcal =	2.5 %
3 % heat loss through insulation	0.13×10^{10} kcal =	3.0 %

Briquetting dryer

450 t of moisture to be evaporated	0.04×10^{10} kcal =	0.9 %
Unused energy	0.11×10^{10} kcal =	2.6 %

II. INVESTMENT COST

Wood Preparation		$ 40,000.-
Carbonizer unit with pyrolysis oil recovery and flare-up chimney (fob plant)		380,000.-
Briquetting plant		400,000.-
Transportation equipment		50,000.-
		$ 870,000.-
Freight, Insurance	$ 70,000.-	
Erecting, start-up	60,000.-	
General cost, contingencies	15,000.-	
	$ 145,000.-	$1,015,000.-
Plant site, preparation	$ 77,000.-	
Start-up	13,000.-	
Working capital	250,000.-	
	$ 340,000.-	
TOTAL INVESTMENT		$1,355,000.-

Plant Site, Buildings, General Services

Cost

Plant site: 8,000 m^2 (incl. preparation, fencing)	$ 10,000.-
Buildings: 800 m^2 (Office, laboratory, warehouse, maintenance)	40,000.-
Hook-up for water, electricity, sewer	12,000.-
Foundations	10,000.-
Water pump, pipes, etc.	5,000.-
Total	$ 77,000.-

III. COST CALCULATIONS

A. Basis

- Waste material free.

- Manpower: 22 including salaried staff, average cost $ 7,000 per year per person.

- Ex-factory price for charcoal briquettes: $ 0.28/kg.

- Ex-factory price for pyrolysis oil (usually 50 % of fuel oil No. 6): $ 0.12/kg.

- Total investment: $ 1,355,000.-

- Depreciation time 15 years: $ 90,000/yr.

- Capital entirely loaned (10 % per year).

- Start up delay: 6 months.

B. Calcualtions

Annual costs:

	Manpowever	$	154,000.-
	Utilities, spare parts,		
	maintenance, lubricants		45,000.-
	Binder, filler, additives		80,000.-
	Bags		36,000.-
	Loan interest		110,000.-
	Subtotal	$	425,000.-
Depreciation			90,000.-
	Total costs	$	515,000.-

IV. ECONOMICS OF THE PROJECT

Annual sales:

	Charcoal briquettes		
	4,500,000 kg x $ 0.28	$	1,260,000.-
	Pyrolysis oil		
	2,800,000 kg x $ 0.12		366,000.-
	Total sales	$	1,626,000.-

1st year of operation (six months)

Sales	$	813,000.-

This is sufficient to cover the operational
costs and interest, resulting in an excess
cash flow of: $ 388,000.-

2nd year of operation (full capacity)

Sales	$ 1,626,000.-
Operational costs	425,000.-
Excess cash flow	$ 1,201,000.-
1st year excess cash flow	388,000.-
Accumulated cash flow	$ 1,589,000.-

This allows the owner to repay loaned capital and still have an excess cash flow of $ 224,000.

3rd year of operation (full capacity)

Sales	$ 1,626,000.-
Operational costs (without interest)	315,000.-
Excess cash flow	$ 1,311,000.-
2nd year excess cash flow	224,000.-
Accumulated cash flow	$ 1,535,000.-

It is estimated that at the end of the third year the return on the invested capital will be higher than 80 %.

Appendix 2 E N E R G Y   D I S T R I B U T I O N   D I A G R A M

This diagram is helpful in selecting the most appropriate carbonization technology.

Since the energy contained in the feed material can be distributed to the end products - char, pyrolysis oil, gas - pilot plant tests and/or tests with bench scale equipment should be conducted.

The energy distribution will depend on various factors, and in particular, the kind of feed. Therefore only the results of efficiently conducted tests will supply reliable forecasts.

By using the diagram on the following page, the charcoal researcher will be able to formulate his decisions in a comparatively short period of time. The basic data needed as input for the diagram should be derived from pilot tests based on a sufficiently large and representative sample of the particular product. These data are:

- Weight yield of the char (% of dry feed).

- Weight yield of the condensable pyrolytic products (% of dry feed).

- Calorific values (kg/cal or kg/KJ) of both the above.

- Calorific value of the dry feed (same units).

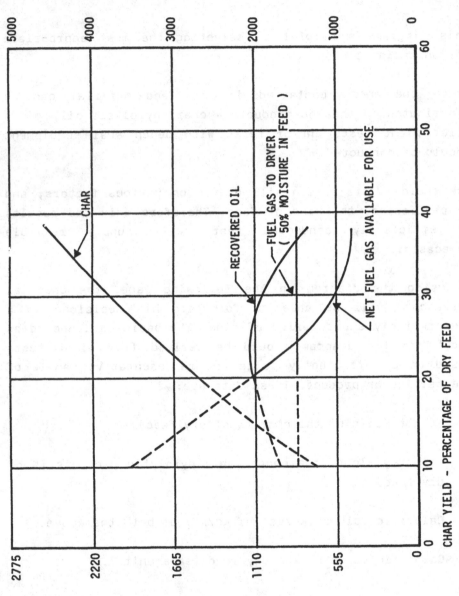

GROSS ENERGY IN PRODUCTS – BTU/LB OF DRY FEED

GROSS ENERGY IN PRODUCTS – kcal/kg OF DRY FEED

CHAR YIELD – PERCENTAGE OF DRY FEED

CHAR

RECOVERED OIL

FUEL GAS TO DRYER
(50% MOISTURE IN FEED)

NET FUEL GAS AVAILABLE FOR USE

(Courtesy of Carbon International, Ltd., Neu - Isenburg , F.R.G.)

Figure 53. Energy distribution diagram

The energy distribution of any carbonization process - solid, liquid, gas - is governed by the specifics of the pure raw material, or raw material mix. It is not possible to forecast these parameters by speculation; however, when one of the numerous possible parameters has been established, and owing to the fact that there is a natural law of the distribution of energy in the plant, the remaining parameters needed for efficient design of a plant may be derived from the diagram.

It is evident that the small charcoal-maker has no need to enter into such costly test procedures.

ADDRESSES OF CONSULTANTS, INSTITUTES, AND EQUIPMENT SUPPLIERS

Note   These lists were compiled to the best ot the author's knowledge, and all names and addresses are given as they were known at the time of writing. The absence of a listing should not be construed as an unfavourable rating because in some cases the publisher was not able to obtain information. Entries are listed in alphabetical order.

CONSULTING FIRMS AND INSTITUTES PROVIDING SERVICES
TO THE CHARCOAL INDUSTRY

1. ALDRED PROCESS PLANT LIMITED
    Oakwood Chemical Works, Sandy Lane
    Worksop, Notts S80 3EY
    United Kingdom
    Primary representative:
    Phone: 0909 476861
    Telex: 54625

2. CARBON INTERNATIONAL, LTD.
    Buchenring 7
    D-6078 Neu-Isenburg 4
    Federal Republic of Germany
    Primary representative: Dr. Walter Emrich
    Phone: 069 - 693201
    Telex: 4 189671 carb d

3. CENTRE NATIONAL D'ETUDES ET D'EXPERIMENTATION
    (CEMAGREE)
    ONEEMA
    Parc de Tourvoie
    92180 Antony
    France
    Primary representative:
    Phone: 666.21.09
    Telex: 204585

4. CENTRE TECHNIQUE FORESTIER TROPICAL
    45 bis, Avenue de la Belle-Gabrielle
    94130 Nogent-sur-Marne
    France
    Primary representative: Jaqueline Doat
    Phone: 873 32 95

5. FOX CONSTRUCTORS & ENGINEERS
    P.O. Box 1528
    Dothan, Alabama 36302
    USA
    Primary representative: H. E. Sprenger
    Phone: 205-794-0701

6. FUNDACAO TROPICAL PESQUISAS E TECNOLOGIA
    Rua Latino Goelho n.o. 1.301
    13.000 Campinas - SP
    Brazil
    Primary representative: Dora Lange
    Phone: 41-7822

7. A. C. Harris
    Consultant
    5 Duncraig Raod
    Applecross
    Western Australia
    Primary representative: A. C. Harris

8. LAMBIOTTE ET Cie. S.A.
    Avenue Brugmann, 290
    B.1180 Bruxelles
    Belgium
    Primary representative: Andre Lecocq
    Phone: (02) 343.01.46
    Telex: 61588 elleco b

9. TROPICAL PRODUCTS INSTITUTE
    Culham. Abingdon,
    Oxfordshire, OX 14 3DA
    United Kingdom
    Phone: 086-730-7551

MACHINERY AND EQUIPMENT SUPPLIERS

1. AEROGLIDE CORPORATION
    P. O. Box Aeroglide
    Raleigh, N. Carolina 27611
    USA
    Phone: 919-851-2000

    Area of activity:
    Wood dryers, charcoal briquetting.

2. ALDRED PROCESS PLANT LIMITED
    Oakwood Chemical Works, Sandy Lane
    Workshop, Notts S80 34Y
    United Kingdom
    Phone: 0909 476861
    Telex: 54625

    Area of activity:
    Portable metal kilns, vertical and
    horizontal carbonizing units.

3. BEPEX, GmbH
    Daimlerstrasse 9
    D-7105 Leingarten
    Federal Republic of Germany
    Phone: 07131-40082
    Telex: 7 28 738

    Area of activity:
    Charcoal briquetting

4. BIO-CARBON, GmbH
    Soecking 26
    D-8254 Isen/OBB
    Federal Republic of Germany
    Phone: 08083-624
    Telex: 526 043

    Area of activity:
    Traditional   and   industrial   carbonization   plants,
    activated   carbon   plants,   training   programmes,
    briquetting.

5. CARBONERA DOCK 80 D
    Enrique Marengo 830
    San Andres, Prov. Bueno Aires
    Argentina

    Area of activity:
    Charcoal briquetting

6. CeCoCo CHUO BOEKE GOSHI KAISHA
    P.O. Box 8
    Ibaraki City, Csaka Prefecture
    567 Japan
    Calbe address: Cecoco Ibaraki, Japan

Area of activity:
Cecoco small-scale kilns, briquetting

7. C. DEILMANN AG-GROUP
    P.O.Box 75
    D-4444 Bad Bentheim
    F.R.G.
    Phone: 05922-72-0
    Telex: 098 833

    Area of activity:
    Pyrolysis, gasification and activation of biomass

8. ENERCO INCORPORATED
    Old Oxford Valley Road #1
    P. O. Box 139 A
    Langhorne, Pennsylvania 19047
    USA
    Phone: 215-493-6565

    Area of activity:
    ENERCO Pyrolytic converter system for biomass conversion

9. ENVIROTECH BSP
    One Davis Drive
    Belmont, California 94002
    USA
    Phone: 415-592-4060
    Telex: 34-5586

    Area of activity:
    Multiple hearth furnaces, large-scale charcoal production

10. ERCO ENERGY RESOURCES COMPANY A.G.
    Zollikofer Strasse 228
    CH-8008 Zürich
    Siwtzerland
    Phone: 01-551010
    Telex: 57-229

    Area of activity: ERCO fluid bed carbonizer

11. LAMBIOTTE ET Cie S.A.
    Avenue Brugmann, 290
    B-1180 Bruxelles
    Belgium
    Phone: (02) 343.01.46
    Telex: 61588
    Area of activity:
    Continuous C.I.S.R. Lamiotte retort

12. LURGI KOHLE & MINERALOELTECHNIK
    Bockenheimer Landstrasse 42
    D-6000 Frankfurt/M. 1
    Federal Republic of Germany
    Phone: 069-71191
    Telex: 4 12360

    Area of activity
    CK-Process (Reichert retort)

13. NICHOLS ENGINEERING & RESEARCH CORP.
    Homestead and Willow Roads
    Belle Mead, New Jersey 08502
    USA

    Area of activity: Multiple hearth furnaces, large-scale
    charcoal production

14. PROCTOR + SCHWARTZ, INC.
    7th Street and Tabor Road
    Philadelphia, Pennsylvania 1920
    USA
    Phone: 215-329-6400

    Area of activity:
    Charcoal briquette dryers

15. PROTRAN INC.
    P.O. Box 10764
    Raleigh, North Carolina 27605
    USA
    Phone: 919-781-4148

    Area of activity:
    Fluid bed carbonizer

16. LA STE CARBOLISI
    Via E. Fermi
    Martara, PV
    Italy

    Area of activity:
    Charcoal plants

17. LA STE LAMBIOTTE - POUR
    Tour Manhattan
    6, place de l'Iris
    92400 Courbevoie
    France

    Area of activity:
    Wood dryers, charcoal plants

Appendix 4   C O N V E R S I O N   T A B L E S

UNITS OF LENGTH

1 mile	= 1760 yards	= 5280 feet
1 kilometer	= 1000 meters	= 0.6214 mile
1 mile	= 1.607 kilometers	
1 foot	= 0.3048 meter	= 30.5 centimeters
1 meter	= 3.2808 feet	= 39.37 inches
1 inch	= 2.54 centimeters	
1 centimeter	= 0.3937 inch	

UNITS OF AREA

1 square mile	= 640 acres	= 2.5899 square kilometers
1 square kilometer	= 1,000,000 square meters	= 0.3861 square mile
1 acre	= 43,560 square feet	
1 square foot	= 144 square inches	= 0.0929 square meter
1 square inch	= 6.452 square centimeters	
1 square meter	= 10.764 square feet	
1 square centimeter	= 0.155 square inch	

UNITS OF VOLUME

1.0 cubic foot	= 1728 cubic inches	= 7.48 US gallon
1.0 British imperial gallon	= 1.2 US gallon	
1.0 cubic meter	= 35.314 cubic feet	= 264.2 US gallon
1.0 liter	= 1000 cubic centi-meters	= 0.2542 US gallon
1 US Barrel	= 42 US gallon	= 34.97 Br. imp. gallon
		= 0.158 cubic meter

UNITS OF WEIGHT

1.0 metric ton   = 1000 kilograms(kg) = 2204.6 pounds (lb)

1.0 kilogram (kg)   = 1000 grams (g)   = 2.2046 pounds (lb)

1.0 short ton   = 2000 pounds (lb)

UNITS OF PRESSURE

1.0 pound per square incl (psi)   = 144 pound per square foot

1.0 pound per square incl (psi)   = 27.7 inches of water[*]

1.0 pound per square inch (psi)   = 2.31 feet of water[*]

1.0 pound per square inch (psi)   = 2.042 inches of mercury[*]

1.0 atmosphere   = 14.7 pounds per square inch (psi)

1.0 atmosphere   = 33.95 feet of water[*]

1.0 foot of water = 0.433 psi   = 62.355 pounds per square foot

1.0 kilogram per square centi-meter   = 14 223 pounds per square inch

1.0 pound per square inch   = 0.0703 kilogram per square centimeter

UNITS OF POWER AND ENERGY

1.0 horsepower (English)   = 746 watt = 0.746 kilowatt (kw)

1.0 horsepower (English)   = 550 foot pounds per second

1.0 horsepower (English)   = 33,000 foot pounds per minute

1.0 kilowatt (kw) = 1000 watt   = 1.34 horsepower (hp) Englisch

1.0 horsepower (hp) (English)   = 1.0139 metric horsepower (cheval-vapeur)

---

*) At 62 degrees Fahrenheit (16.6 degrees Celsius)

1.0 metric horsepower	= 75 meter X kilogram/ second
1.0 metric horsepower	= 0.736 kilowatt = 736 watt
1 kilowatt hour	= 3.412 British thermal units (Btu)
	= 1.34 horsepower hours
	= 3,600 kilojoules
	= 3.6 megajoules
1 British thermal unit (Btu)	= 1055,2 Joules (J)
	= 0,252 kilo calories (kcal)

FUEL CONVERSIONS

(a) 1 quad = 1×10^{15} Btu (quadrillion Btu)

1 quad = 40×10^6 tons bituminous coal;

$\quad\quad\quad = 50 \times 10^6$ tons sub-bituminous coal; and

$\quad\quad\quad = 62.5 \times 10^6$ tons lignite

1 quad = 172.4×10^6 barrels of oil

1 quad = 1×10^{12} ft^3 natural gas

1 quad = 62.5×10^6 tons wood (O.D. basis)

1 quad = 96.2×10^6 tons wood (green basis)

1 quad = 105×10^6 tons municipal waste

1 quad = 293×10^9 KWh delivered

(b) 1 ton bituninous coal = 25×10^6 Btu

(c) 1 barrel of oil = 5.8×10^6 Btu

(d) 1 ft^3 natural gas = 1000 Btu

(e) 1 ton wood (O.D. basis) = 16×10^6 Btu[*)]

$\quad$ 1 ton wood (green basis) = 10.5×10^6 Btu[*)]

(f) 1 ton municipal waste = 9.5×10^6 Btu[*)]

(g) 1 KWh (delivered) = 3412 Btu

[*)] Average value, subject to wide variation

WOOD CONVERSION

(a) 1 cubic ft = 30.0 lb[**]

(b) 1 cord = 3,62 m^3 = 1,25 ton

(c) 1 bd ft = 2.5 lb[++]

(d) 1 stere = 1 cubic meter (cube with edges of 1 m length)
    = 510 kg for pilewood (Europ. hardwood)[+]

[**] 1 ft^3 softwood = 27 lb and 1 ft^3 hardwood = 32 lb on average.

[+] A cord USA is defined as wood stacked in a 4 x 4 x 8 ft pile. There is much variation in this unit of measure. Because of wood density, free space within a cube made of stackwood, weight is subject to wide variations

[++] Approximate value.